Physics for Poets

Physics for Poets

FOURTH EDITION

Robert H. March

Department of Physics and Integrated Liberal Studies
University of Wisconsin–Madison

McGraw-Hill, Inc.

New York   St. Louis   San Francisco   Auckland   Bogotá   Caracas
Lisbon   London   Madrid   Mexico City   Milan   Montreal   New Delhi
San Juan   Singapore   Sydney   Tokyo   Toronto

This book was set in Palatino by the Clarinda Company.
The editors were Jack Shira and James W. Bradley;
the production supervisor was Elizabeth J. Strange.
The cover was designed by Lisa Cicchetti.
The photo editor was Anne Manning.
R. R. Donnelley & Sons Company was printer and binder.

Cover Photo
© Shoji Sato / Photonica

PHYSICS FOR POETS

This book is printed on acid-free paper.

2 3 4 5 6 7 8 9 0   DOC DOC   9 0 9 8 7 6

ISBN 0-07-040248-5

Library of Congress Cataloging-in-Publication Data

March, Robert H., (date).
 Physics for poets / Robert H. March.—4th ed.
  p.   cm.
 Includes index.
 ISBN 0-07-040248-5
 1. Physics.   I. Title.
 QC23. M334   1996
 530—dc20   95-18959

About the Author

ROBERT H. MARCH has been on the faculty of the University of Wisconsin in Madison, Wisconsin, since 1962. He has taught in the physics department and in an interdisciplinary program called Integrated Liberal Studies. He was educated at the University of Chicago, where he began his scientific career as a technician in the laboratory of Enrico Fermi. He has done experimental research in particle physics, the subject matter of Chapter 19 of this text, at laboratories in the United States and in Europe. Since 1980 his research interests have shifted to astrophysics, using facilities located in the Hawaiian Islands.

*To the memory of Professor Gretchen Schoff,
whose life and work personified everything
this book stands for*

Contents

Eighteen   Schrödinger's Cat   225
Nineteen   The Dreams Stuff Is Made of   234

Preface

The author of this volume is a member of that favored generation that emerged from graduate school around 1960. The post-*Sputnik* explosion of federal research support coincided with the arrival at college of the first baby boomers, and with the euphoric optimism of the Kennedy era, to generate opportunities that would be unimaginable in today's more austere climate. Those of us who chose particle physics were doubly blessed with participation in the exciting period of discovery that culminated in the so-called Standard Model.

However propitious those years may have been for our careers, they proved far less so for the education of our students. Teaching was all too often overwhelmed by the demands of research. The early editions of this book were motivated by a desire to restore some equilibrium by offering professors the opportunity to teach the twentieth-century science they knew and loved, and the nonscience student some exposure to the revolutionary ideas on which that science was built. It was inspired by a conviction that, properly viewed and properly presented, physics was a part of the humanistic as well as the scientific tradition.

Today, with the demise of the Cold War as the focus around which contemporary Western society organized itself for more than four decades, the motivation for public support of fundamental research is no longer clear. More than ever, it behooves us to show yet another generation of students the human face of science.

It should also be noted that the Standard Model, however successful it may have been at explaining the fleeting phenomena of the submicroworld, falls far short of the reductionist dream of stark simplicity on nature's most fundamental level. The final chapter of this work attests to that. Those who still dream that dream, which has animated physical thought for most of its modern history, seem to have entered a regime of diminishing returns on their intellectual labors.

Whether this proves merely a temporary pause, or presages a major historical turning point, it is clear that much of the current science that is capable of exciting the young minds that this book is designed to reach moves along other lines. To borrow a term, these might be characterized as "C-cubed," for cosmology, chaos, and complexity. This edition introduces these topics in a modest fashion appropriate to its limited scope.

Another change from recent editions is a minor upgrade in the mathematical level of the presentation, in recognition that the students currently taking this course are a bit more comfortable with numbers (but, alas, far less comfortable with history) than their predecessors.

The author would like to thank his colleagues and students in the Integrated Liberal Studies Program and the Bradley Residential Learning Community for stimulating discussions and continuing inspiration.

McGraw-Hill and the author would like to thank the following reviewers for their many helpful *comments and suggestions:* Ernest Abers, University of California, Los Angeles; David Atkatz, Skidmore College; Robert Austin, Princeton University; Howard Brody, University of Pennsylvania; Harry A. Brown, University of Missouri; Ira Cohen, Linfield College; Donald Johnson, St. Thomas Acquinas College; David Kammer, Albion College; Rene Ong, University of Chicago; Joseph Schick, Villanova University; Sally Seidel, University of New Mexico; Joseph Spizuoco, Shippensburg University; Erick Weinberg, Columbia University; and Carol Zimmerman, Tallahassee Community College.

Robert H. March

Physics for Poets

Introduction

To the laboratory then I went. What little
right men they were exactly! Magicians
of the microsecond precisely wired
to what they cared to ask no questions of
but such as their computers clicked and hummed.

It was a white-smocked, glass, and lighted Hell.
And there Saint Particle the Septic sat
lost in his horn-rimmed thoughts. A gentlest pose.
But in the frame of one lens as I passed
I saw an ogre's eye leap from his face.
—JOHN CIARDI, Fragment

A book entitled *Physics for Poets* may fittingly be opened with a poem. The one above is introduced as Exhibit A, evidence of the need for a book entitled *Physics for Poets*. John Ciardi was entitled to his own personal vision, to be sure, and this one is not entirely without foundation. But it seems sad that a poet who lived much of his life a short drive from the home of Albert Einstein missed seeing another side of science. Like poetry, science is a creative activity that engages the emotions as well as the intellect, and its best practitioners deserve to be called *artists*.

A scientist, of course, is supposed to be looking for the *truth* about nature. But not all truths are equal. Some we call *deep* truths, and these are the ones that are also *beautiful*. An idea must be more than right—it must also be pretty if it is to create much excitement in the world of science. For the search for truth is not simply a matter of discovering facts. You must also understand their significance, and then persuade others that your way of looking at them is valid. It is always easier to persuade people to believe in something new when they find it beautiful, especially if it runs counter to their established beliefs.

What, then, is the meaning of *scientific objectivity?* It does *not* mean that a scientist must be cold and dispassionate. Science is a combat of ideas, and objectivity simply means to fight fair. A scientist may believe–*passionately*–in his or her own point of view. But contrary ideas must be given due consideration, not dismissed without mention. Embarrassing facts must be confronted, not ignored.

Science is supposed to be a great adventure. Well it may be; but the student approaching a conventional introductory science course with this maxim in mind may be in for a letdown. These courses are usually tailored to the

1

needs of future scientists and technicians, and concentrate on developing the skills and terminology such people need. The fun stuff will come in more advanced courses. Rarely does much of the sense of adventure manage to come through the hard work and tedium.

But one wonders how many people would love music if everyone was required to practice scales on a piano for a certain number of hours before being allowed to listen to a piano sonata. True, a concert pianist probably enjoys the sonata on some levels denied to others, but a reasonably sensitive person with totally untrained fingers can appreciate its beauty. And the analogy with music may not be as farfetched as it seems. To carry it further, this book will let you listen to a bit of Bach and then see how you make out with Stravinsky and Bartók.

Physics has seen two periods of rapid change in its most fundamental ideas. *Revolution* is one of the most abused words in the English language, but in these cases it is probably appropriate. The first revolution occupied most of the seventeenth century and was so complete that little that preceded it can be recognized as physics at all, in modern terms. The second has taken most of the twentieth century, and we have clearly not seen the end of it.

It is convenient to regard the first revolution as beginning with Galileo and culminating with Newton (with some injustice to many worthwhile predecessors and contemporaries of these two heroes). It created classical mechanics, probably the most successful scientific theory of all time. For two centuries, this theory swept all before it, one phenomenon after another yielding to explanation in mechanical terms. By the end of the nineteenth century, it seemed on the verge of absorbing all of science.

Indeed, to many scientists of that time, Newtonian physics appeared already to have done so, except for a few minor details. Yet, as a result of the effort to account for these details it was not just modified, but essentially abandoned. Just how such a successful intellectual edifice, erected on an immense and solid base of fact, can be replaced by something dramatically different, is a central theme of this book.

The triumph of Newton's mechanics had wide repercussions. Leaving aside the legion of (mercifully) forgotten Deist theologians who came to look upon the Creator as a sort of master clockmaker, we can trace its impact on nearly all aspects of Western culture. To many intellectual leaders Newton's physics became a model that all of human knowledge should strive to emulate. This was unfortunate, for much of the power of grand theories comes from ignoring the messy but important details that rule our practical, day-to-day lives.

The effect has been particularly pernicious in the field of economics, where the Newtonian model dominates to this day. Modern physics has taught us that no matter how clever our theories, how careful our measurements, or how powerful our computers, the future must always remain largely unpredictable.

The second revolution has not yet produced anything like the comfortable unity of Newton's system, for it struck out in two very different new directions: relativity and quantum theory. Relativity was largely the creation of one man, Albert Einstein. Quantum theory grew from the contributions of many (including Einstein). Relativity is popularly regarded as bizarre and abstruse, but quantum theory is, in many ways, far more so.

The realm of the quantum theory is the very small, while relativity deals with the very large or the very fast. Where they come together, in the very small very fast world of elementary particles, they have not yet achieved a harmonious unity. The key unresolved problem is the oldest one in physics, the one that launched Galileo's revolution, that of *gravity*. It can truly be said that this remains, as it always has been, the central mystery of our physical universe.

Both theories were conceived, at least in part, in much the same spirit–that of critical evaluation of the process by which we actually *observe* the world. Both deal mainly with phenomena that lie outside the realm of ordinary experience. It is partly for this reason that they are so difficult to teach–the phenomena themselves are far removed from anything we can see, touch, or feel.

The same, however, could have been said of Newton's physics, which won its spurs by accounting for the motions of the planets, a problem far removed from the concerns of home, hearth, and the workbench. Most of us approach practical matters through commonsense precepts that can only be regarded as *pre*-Newtonian.

In our century, science has become a large and well-organized profession. It is this professionalism that has made so much of science inaccessible to the general public. In their working lives, scientists need only communicate with fellow specialists. To communicate with "outsiders" takes time and effort, and is not likely to advance an academic career. Even when they make the effort, many scientists have no idea what to say. This is because it is perfectly possible to contribute to science without thinking too hard about the deeper significance of what one is doing.

In physics, the most serious barrier to communication is mathematics. Some of the beauty of this science is most readily apparent only when it is written in its native tongue, which is partly mathematical. A lot of this beauty is unavoidably lost in translation. To ask someone to study mathematics merely to appreciate physics is as unreasonable (or as reasonable) as asking them to study Italian merely to properly appreciate Dante. Of course, like Italian, mathematics is beautiful in itself and is likely to be useful for a variety of other purposes.

In this book mathematics is kept to a minimum, the criterion being that the effort must make a point, or illuminate a concept, in an indispensable way. It never goes beyond arithmetic and first-year high school algebra, except in a few places where, for the benefit of the mathematically adept, something more advanced is put on display. These sections are clearly designated, and may be skipped without losing the continuity of the story. In some cases, the mathematics is relegated to footnotes. Simply regard these cases as akin to the brief excerpts from the original that are occasionally inserted in a literary translation.

The worst possible attitude with which to approach the study of any science is one of awe. Like most successful human ventures, science prospers largely by sticking strictly to business. There are problems that lend themselves to its methods, and the solution of them can enrich the human experience, both materially and intellectually. But abstract science, powerful as it may be in its own domain, is neither universal nor magic. Most of what humanity holds dear must remain beyond its scope. If science has achieved much, it has been by limiting its reach to those things that most easily fall within its grasp.

People raised in Western cultures are taught that science is the place to go when you want hard, incontestable truths. This is unfortunate because, if anything, science shows what an elusive commodity truth can be. Despite all professional training and care, observations are often in error, and even more often their implications are misinterpreted or overlooked. Even a simple report of a fact can bear the stamp of a theoretical framework. A scientist who has never had to give up some cherished precept has probably been working in a pretty dead area. It is this experience of *having been wrong*, rather than any smug conviction of always being right, that characterizes the scientific outlook.

What science *should* teach us is to *doubt*—to consider that many of the beliefs we take for granted may arise from custom, or prejudice, or may simply be *wrong*. There is no more important lesson that a citizen of our troubled world can learn, and if this book can contribute to teaching that lesson, it will have done its job.

CHAPTER 1

A Vast and Most Excellent Science

The Book of Nature is written in mathematical characters.
—GALILEO GALILEI, The Assayer

A revolution must have its heroes, and the one that gave birth to modern physics was blessed with two of truly mythic stature, Galileo Galilei and Isaac Newton. In the heroic legend, Galileo is cast as the Martyr of Truth, and Newton as the Saintly Hermit of Reason. In the flesh and blood, they were considerably more complicated—and more interesting—than these figures of myth.

Galileo's most enduring legacy to science was, if anything, the *least* controversial product of a life filled with controversy. His *Discourses and Mathematical Proofs Concerning Two New Sciences,* usually referred to simply as *Two New Sciences,* was published in 1636. By then the author was in his seventies with failing vision, living under a genteel sort of house arrest. The book was largely a summary of work done decades before, when he had been a young professor of mathematics at the University of Padua.

Forbidden by order of the Church to claim that the Earth was moving, Galileo chose a more devious stratagem—to undermine the old physics that led people to believe that it *couldn't* move. When he made the seemingly innocent observation that something dropped from the top of the mast of a moving ship will land at the base, Galileo was speaking in a kind of code. He was really trying to persuade his readers that the Earth could be hurtling through space without our being the least aware of its motion.

The centerpiece of the book is a mathematical description of how things move as they fall. It was a model for a new kind of science. According to the myth, this was a science more faithful to the evidence of the eye than the dogmas it would replace. But Galileo was not all that concerned with the real world. His new science was rooted in an abstraction, an imaginary world without any air to complicate the motion.

For Galileo's greatest gift to his successors was to liberate his infant science from the obligation to deal with the full, confusing hurly-burly of mundane reality. Unfettered, physics was free to move ahead. Half a century later, the stratagem of Two New Sciences bore fruit in the work of Isaac Newton, whose law of gravity finally convinced the world that the Earth really does move.

A RENAISSANCE MAN

Galileo lived from 1564, the year of Shakespeare's birth and Michelangelo's death, until 1642, the year Newton was born. He was the eldest son in a respected Pisan family of modest means. His father, a musician and scholar who wrote one of the first modern treatises on harmony, hoped his clever son might recoup the family's fortunes by means of a medical career. But at the university, Galileo became fascinated with mathematics, then as now a far more precarious way to earn a living.

Even Galileo's greatest admirers must admit that he could be boorish, pugnacious, and petty. He was sometimes unscrupulous in his ambition, on several occasions claiming credit for the work of others. He had a remarkable gift for the written word and could not resist the temptation to sprinkle his works with elegant insults to his rivals. In short, he was very much the late Renaissance man, a prodigy who could paint or versify as well as solve an equation.

By the age of 25, Galileo's writings and inventions had made him enough of a reputation to land the poorly paid position of lecturer in mathematics at the University of Pisa. Within three years he moved on to a more prestigious post, a professorship at the University of Padua. This university and its neighbor in Bologna were the two oldest in Italy, and were generally regarded as the best in the world. With such an affiliation, he could become someone to reckon with.

Padua was part of the Republic of Venice, which was dominated by hard-nosed merchants. They were proud of their great university, but found it prudent to let professors fend for themselves financially. Galileo was obliged to help out a younger brother and sister, as well as provide for his own small family, and the fees from the few students of mathematics who came his way were never enough. He supplemented these by his writing and from the sale of measuring instruments of his own improved design.

A major turning point in his life came at the age of 45, when he developed the astronomical telescope, his refinement of a novelty invented by the eye-glass makers of Holland. In a celebrated pamphlet, *The Starry Messenger*, he reported the wonders it revealed: the moons of Jupiter, mountains on our own moon, and the phases of Venus. These discoveries were dedicated to an illustrious and powerful former student, Cosimo di Medici, the young grand duke of Tuscany.

As Galileo had hoped, Cosimo was flattered to the point of opening his family's bountiful coffers. Galileo was invited to become "Chief Philosopher and Mathematician" to the celebrated Medici court. He would serve as a sort of resident wise man who could offer technical advice, cast an occasional horoscope, and participate in debates that were a common after-dinner diversion for late Renaissance princes and their guests. But most of all he was there to lend some up-to-date intellectual glitter to the artistic splendor of a city that, even in decline, was still a wonder of the world.

With a patron as wealthy as Cosimo, Galileo enjoyed not only an incomparably higher standard of living than his professorship could provide, but the opportunity to reach a wider audience that included the most influential leaders of his time. But all of this comfort and acclaim carried a price tag. In Padua, Galileo had been somewhat insulated from the political and religious infighting of his day. In the palace life of Florence, his position was both more public and more vulnerable, and his sharp tongue soon earned him powerful enemies.

A demonstration of his wondrous instrument before the papal court in Rome was greeted with enthusiasm. This emboldened him to claim that the things the telescope made visible were incontrovertible proof of Nicolaus Copernicus' claim that the Earth and the planets all revolved about the Sun. Throughout the half century since this idea first appeared in print, the Church had taken no official stand on the question. Galileo had kept his own opinions private. Going public proved as rash as it was impulsive.

In retrospect, it is easy to see why his arguments were unconvincing, even to the many people sympathetic to his point of view. No observation of the skies can tell you whether the Earth is moving or standing still. In Chapter 4, we will see that what was required to settle the question was some new physics.

Galileo had made himself a sitting duck for his enemies, who brought the dispute before Church authorities. But these were not the prelates of a century before, whose haughty disdain for public opinion had brought forth Martin Luther's ninety-five theses. Though Copernicus' great work was banned, Galileo himself escaped censure. He was simply warned that while he could continue to discuss the Copernican view as a "hypothesis," he must stop claiming to have established it as an indisputable fact.

With the accession to the papal throne of his old friend Cardinal Barberini, who became Pope Urban VIII in 1623, Galileo saw a chance to cap his career in a book that would persuade the Church to tolerate, or even to adopt, the Copernican system. Urban was receptive, but proposed his own formulation of how to establish a peaceful harmony between science and theology: we poor humans cannot presume to know how things *really* are—that is reserved for an all-knowing God. Science can at best improve the description of how things *appear* to be. Accepting this formula, Galileo was allowed to publish, in 1632, the *Dialogue on the Two Great World Systems.*

The work was widely acclaimed from the day of publication, and the tone was Galileo at his devious best. It was cast as a Platonic dialogue, like the script of a play. Arguments for both sides were there, to be sure, but the Earth-centered view was defended by a character named Simplicio, who was portrayed as a dull-witted, pedantic nitpicker. And Galileo was not content to rest his case on the book alone. At social gatherings in Florence and Rome, he would regale the guests by taunting his adversaries and mocking their views.

Furthermore, like the *Starry Messenger,* this new book was not completely convincing on its most controversial point. Most objective readers were convinced by its arguments that the planetary motions were centered on the Sun. What they would not buy was the notion that the Sun must therefore be standing still and the Earth moving. This flew in the face not only of intuition, but of the established physics of the time. Without a new physics to replace it, a prudent astronomer, as well as a scientific layperson, could be forgiven a bit of skepticism over Galileo's claim to have established the reality of the Earth's motion.

Pope Urban felt flagrantly double-crossed, even by the loose standards of that age. In trouble himself because of the failure of military adventures, he decided that this time Galileo had gone too far. Urban could be a patient man, willing to give considerable latitude to scientific debate, but the Church must at least be allowed to make the rules. A special tribunal summoned Galileo to Rome, where he was forced to renounce his opinion that the Earth moved. He was never officially condemned, but was removed from public life.

The Church, however, had put itself in a no-win situation. In order to assert its authority, it had to swim against the tide of history. The Renaissance had elevated science to a high status. The printing press had opened the way to wide, rapid dissemination of new ideas, ending the monopoly on learning of the academies and monasteries. A worldly humanism, with a growing faith in the power of man's reason, was the spirit of the time.

Galileo played this game well, writing in Italian rather than scholarly Latin, to take his case to the growing educated lay public instead of confining it to polite discussion in academic circles. This further infuriated his enemies.

But he was not playing a lone hand. Galileo was a member of the Accademia dei Lincei (Academy of the Lynxes), which was something between a learned society and a secret social club of the young nobles of Rome. It met to dine and debate, fostered scientific correspondence, and helped members and their protégés get their works in print. They were the wave of the future, and both they and the more perceptive leaders of the Church knew it.

GALILEO TAKES ON ARISTOTLE

The central intellectual event of the early Renaissance had been the rediscovery of classical Greek philosophy. The Greek intellectual heritage had been preserved through the Middle Ages more by the world of Islam than by anyone in Christian Europe, which was pretty much a barbarous backwater right up to the thirteenth century. The ancient philosophers had not been entirely forgotten, but only fragments of their writings were available in Europe, and many of these were kept under lock and key lest they infect the reader with the pagan religion of their authors.

Algebra and a bit of trigonometry, both Arab embellishments of Greek and Indian achievements, represented the pinnacle of mathematical knowledge. If your own mathematical education stops short of calculus, most of what you know was taught in Moslem universities from Toledo to Timbuktu before the fourteenth century.

In the early stages of Europe's recovery, Thomas Aquinas and others molded ideas taken from Greek thought, notably the work of Aristotle, into an all-encompassing worldview known as *scholasticism*. This system was most highly developed in areas that connected to moral philosophy and theology, adding little to Greek and Arab science. Mathematics and astronomy were respected as exercises to discipline the mind, but otherwise these profane sciences did not count for much.

The unifying principle of scholasticism was *teleology*, the study of the *purposes* of things. For readers familiar with the Aristotelian system, this is what he called the "final cause" of anything. It could be something quite mundane: the final cause of a chair is "to be sat on." But in the context of natural science, it acquires a religious significance. The philosopher observes nature in order to discern God's will at work, thus providing a harmonious link between science and theology.

It is unfair to accuse scholastic scholars of disdain for the evidence of their own eyes. Aristotle himself had been a first-rate biologist. One-fifth of his surviving writings are on this topic, and for thoroughness of observation and clarity of insight they set the standard for centuries to come. He was a careful, systematic observer who believed in an order that resided in nature, waiting to be uncovered through observation, comparison, and classification. The great descriptive sciences, such as geology and zoology, were primarily Aristotelian in their methods right up to the middle of the twentieth century.

Galileo's approach was a bit more sophisticated. His Aristotelian schooling had taught him to respect the value of observation. But he had even greater admiration for Aristotle's teacher Plato, who had worshipped the sublime abstract beauty of geometry. Brute nature offered nothing as perfect as the ideal triangle of dimensionless points and lines. What we scrawl on paper can be only a feeble approximation. Yet that wondrous perfect triangle exists only in the human mind, visible only to pure reason. We must look beyond the imperfect world revealed by our fallible senses to uncover a higher, more perfect reality, the only fit object of study.

Galileo found a happy fusion of these contrasting approaches. Like Plato, he quested after hidden truth, written in the deep language of mathematics. But experience had shown him that reason, unaided by the senses, can easily be led astray. Passive observation, however, is no better, for nature is too sly an adversary to reveal her most treasured secrets to any fool. You must confront her armed with the best instruments human ingenuity can devise. Even then, she cannot be taken on her own ground. She must be tricked into showing her hand by contriving situations that emphasize the hidden reality. This is the essence of experimental science, and Galileo's notebooks reveal that he practiced it with a master's hand.

But above all, Galileo rejected teleology. It is labor enough to uncover the *how* of nature; the *why* must remain forever beyond the methods of science. To this day, teleological arguments are strictly off limits in science, although privately some scientists may believe that they discern a purpose behind the workings of nature.

Aristotle had devised a rule for dealing with motions that continue at constant speed. In the heavens, the stars move in this fashion, wheeling about the Earth like points of light in a giant sphere, turning slowly and evenly. On Earth, ships and horse-drawn carts also maintain a fairly constant speed for long periods of time, so this was of practical interest as well.

Aristotle saw this kind of motion as a contest between *propulsion* and *resistance*. The resulting speed, he guessed, would depend on the ratio between the two. Double the effort, by adding rowers or sails, and a boat might well move twice as fast. Cutting the resistance was another way to add speed, which was why the wheel was invented, and why boatbuilders carefully smoothed their hulls. Though Aristotle lacked any means of putting his rule to a quantitative test, we will see in Chapter 5 that he was not far from the mark.

Even Aristotle realized, however, that there was one familiar form of motion that was hard to explain within this scheme. Falling objects quite obviously pick up speed as they drop. What is happening here? Is the resistance decreasing, or is the propulsion growing? Over the centuries, Aristotelians debated the point, but could offer no ready answer.

There was another embarrassing problem. If it is the weight of an object that propels its fall, shouldn't heavy objects fall faster than lighter ones? Indeed, a literal reading of Aristotle would suggest a speed of descent that was fully in proportion to weight, while experience teaches us that light objects fall nearly as fast as heavy ones. Though Aristotle never thought of his science in terms of immutable laws, but simply as generalizations from experience that might allow for some exceptions, Galileo chose this weak point to bear the thrust of his assault.

In a key passage from *Two New Sciences,* cast like the banned *Two World Systems* as a Platonic dialogue, the character Salviati is the author's mouthpiece. The hapless Simplicio is clearly no match for him, and can only fall back on the authority of the written word. The third interlocutor, Sagredo, represents the intelligent, pragmatic humanist that Galileo hopes to win to his side.

SALVIATI: I greatly doubt that Aristotle ever tested by experiment whether it be true that two stones, one weighing ten times as much as the other, if allowed to fall, at the same instant, from a height of say, 100 cubits, would so differ in speed that when the heavier had reached the ground the other would not have fallen more than 10 cubits.

SIMPLICIO: His language would seem to indicate that he had performed the experiment, because he says: "We see the heavier": now the word *see* shows that he had made the experiment.

SAGREDO: But I, Simplicio, who have made the test can assure you that a cannon ball weighing one or two hundred pounds, or more, will not reach the ground by as much as a span ahead of a musket ball weighing only half a pound. . . .

However, tearing down is always easier than building up, and there had been many critics of Aristotle. Galileo earned his present place in scientific esteem by offering an alternative description of the motion of falling bodies, in two succinct statements:

1. In a medium totally devoid of resistance all bodies will fall at the same speed.
2. During equal intervals of time a falling body receives equal increments of speed.

The words *totally devoid of resistance* may seem innocent enough, but they are in fact words of defiance. First of all, they are a clear break with the Aristotelian formulation, in which zero resistance would imply an infinite velocity. They also suggest a *vacuum*, which the prevailing scientific thought considered a most unnatural state: "nature abhors a vacuum." Finally, since a vacuum was then impossible in practice, the words proclaimed Galileo's refusal to take nature as he found it. It was his intent to describe a state of perfection to which nature can only approximate.

Nonetheless, Galileo could not cavalierly dismiss Simplicio's objections. Feathers do indeed fall more slowly than cannonballs. Here the master of experiment asserts himself. If he cannot make resistance *go away*, he will prove his point by *making it worse!* Salviati reports the result:

> Have you not observed that two bodies which fall in water, one with a speed a hundred times as great as that of the other, will fall in air with speeds so nearly equal that one will not surpass the other by as much as a hundredth part? Thus, for example, an egg made of marble will descend in water one hundred times more rapidly than a hen's egg, while in air falling from a height of twenty cubits, the one will fall short of the other by less than four finger-breadths.

In short, if the deviations from his law are far worse in a dense medium than in a thin one, is it not reasonable to suppose that they would disappear if the medium were absent altogether?

As a final touch, Galileo insisted that the question of the possibility or impossibility of a vacuum was quite irrelevant to the validity of his law. It is possible to understand nature in terms of approximation to an ideal state, even if that state cannot possibly exist in nature.

The second part of Galileo's description—how falling bodies gain speed—was not a particularly radical departure. Aristotle might well have applauded it as an extension of his own style of physical thinking. But it does serve to illustrate the two roles that mathematics would have in the "new science." First of all, it would extend our ordinary language. Second, it would provide a means of producing quantitative predictions that would be the acid test of his ideas.

THE MATHEMATICAL LANGUAGE OF MOTION

The word *speed* is familiar, and so is its mathematical definition:

$$\text{speed} = \frac{\text{distance moved}}{\text{elapsed time}}$$

The word *velocity* is a synonym for speed in ordinary English. As used by physicists, it has a slightly different meaning, as will be explained in Chapter 2.

In order to deal with things whose speed changes, such as Galileo's falling bodies, we should qualify the above definition by calling it *average* speed.* The

---

*The algebraic definition of average speed is $\bar{v} = \Delta x / \Delta t$. The bar over the v designates an average value.

Portrait of Galileo by Sustermans.
(The Granger Collection.)

speed at any instant can be measured by choosing a time interval that is short enough. Isaac Newton's calculus is a way of describing things by imagining *infinitesimal* intervals, but we will not need to go that far. The simple cases we will treat may be visualized with the aid of a graph of distance vs. time.

When something moves at constant speed, the graph is a straight line, as in the two examples in Figure 1-1. The steeper the slope of this line, the greater the speed. If the speed is changing, the slope of the line must change. In that case, we get a curved line, like the one in Figure 1-2. When we see a graph like this, we describe the motion as *accelerated.*

In everyday speech, acceleration means just one thing, speeding up. As a physicist uses the word, however, it can also stand for slowing down. In that case, the acceleration is *negative.* In Chapter 3, we will see this definition extended even further, nearly losing contact with common usage. This tendency to borrow words from ordinary language, and then modify their meanings to suit a mathematical definition, is one of the things that helps make physics confusing.

The mathematical definition of *acceleration* is contained in the formula

$$\text{acceleration} = \frac{\text{change in speed}}{\text{lapsed time}}$$

FIGURE 1-1. Graphs of motion at constant speed.

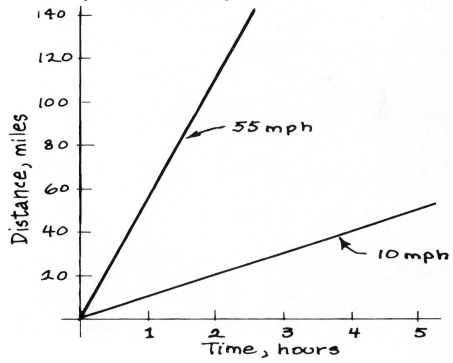

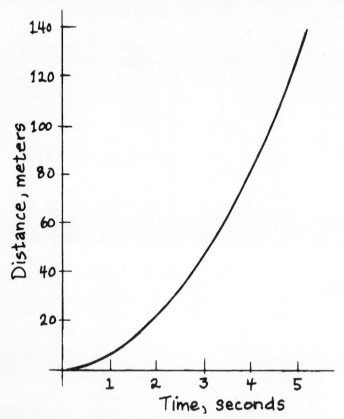

FIGURE 1-2 Graph of uniformly accelerated motion.

For example, a car that can go from zero to 45 miles per hour in 5 seconds experiences an acceleration of $45/5 = 9$ miles per hour per second.

If we allow acceleration to be a negative quantity, it serves equally well to describe slowing down. For example, if a car slows from 50 to 20 miles per hour in 6 seconds, the change in speed is $20 - 50 = -30$ miles per hour, and thus the acceleration is $a = -30/6 = -5$ miles per hour per second.

In these examples, we have deliberately chosen different time units for specifying speed (hours) and acceleration (seconds). Physicists usually use the same units for both. In their favorite system of measures, the SI or "metric" system, that 9 miles per hour per second becomes 4 meters per second per second, which is then shortened to meters per second squared or, in abbreviation, m/s^2. The term can be a source of confusion—what on Earth is a "square second"? The answer is that the square merely signifies that to get from measurements of distance to those of acceleration you must divide by time *twice*.

Had Galileo's readers been more familiar with this language, he might have condensed his description of falling body motion to a single succinct statement:

In the absence of resistance, all falling bodies experience the same constant acceleration.

If the car in our example gains speed in that fashion, at the end of one second it will hit 9 mph, 18 mph at the end of two, and after ten seconds would be at 90 mph. Similarly, if it slows down at a constant −5 mph/s it will take ten seconds to come to a dead halt from 50 mph.

But in Galileo's day, there were no instruments that could directly measure either acceleration or speed. Here is where a second role of mathematics comes in. Galileo's law was stated in terms of something that could not be directly observed. Mathematics would now derive from it a statement about things that *could* be. With the help of a bit of simple algebra, Galileo was able to convert his statement about constant acceleration into a relation between the things he could measure, *distance* and *time*.

The relation he derived was that for an object accelerating uniformly from rest, the distance was proportional to the square of the time.*

For example, an object released from rest will drop about 5 meters in the first second. Then it will fall four times as far, or 20 meters, in two seconds, 45 meters in three seconds, and so on.

The reason for squaring the time is straightforward. To go from acceleration to speed, multiply by time. To get from speed to distance, multiply by time again, and you have time squared.

Getting down to measurables was not the end of Galileo's difficulties. With the best time-measuring instruments of his era, he could scarcely measure intervals to a fraction of a second. Yet a heavy object dropped from the highest towers available to him would be in flight for little more than three seconds. He had to find a way to stretch out the time.

To remove this difficulty, Galileo chose *not* to study freely falling objects. Instead, he made measurements on a ball rolling down an inclined plane. He asserted (but could not actually prove) that this would "dilute" the motion (i.e., reduce the acceleration) without fundamentally altering its character.

Using a smooth board with a small tilt and a groove to guide the ball, Galileo was able to study a motion that took about ten seconds to complete. His timer was a water jar with a spigot at the base. The ball was released from various positions along the board, and while the ball rolled the spigot was opened and water flowed into a cup. Afterward, the cup was weighed, and the weight taken as a measure of the time. The distance traveled was in proportion to the square of the weight of the water, bearing out Galileo's prediction.

Here is the essence of Galileo's experimental method. Start with an idealized description, stated in terms of unmeasurable quantities. Use your math-

---

*The algebraic expression of this rule is $x = \frac{1}{2} at^2$, where x stands for distance, a for acceleration, and t for time. The $\frac{1}{2}$ is there because the average speed of an object accelerating from rest is half its final speed.

ematical skill to convert that into a statement about things you can measure. Does nature move too quickly for you to follow? Slow her down in a way that you think changes nothing important. This is a far cry from the Aristotelian scientist out in the field, taking nature as given and carefully observing and sketching in a notebook.

But neither was it pure Platonism. An effort to understand free fall from that perspective had been made a generation earlier, by none less than the quintessential Renaissance man, Leonardo da Vinci. Searching for a simple mathematical rule, Leonardo guessed that the distances traveled in successive seconds would follow the sequence of ordinary numbers, e.g., 1, 2, 3, 4, ... This was pure numerology, and it was not quite right.

What gave Galileo the edge over Leonardo? One asset was of course his mathematical training. But it is perhaps more significant that he had started knowing the correct relation of distance to time, having already discovered it experimentally! The principle of uniform acceleration was not his starting point, but the conclusion of a long chain of discovery, as we shall see in the next chapter.

WAS ARISTOTLE SO WRONG AFTER ALL?

If pure reason can lead a scientist astray, so too can unreasoned experimentation. If Galileo had been able to observe the fall of fairly light objects from great heights, and present the results in graphical form, he would have obtained the curve shown in Figure 1-3. The reason for this curious behavior is very simple: As a body speeds up, the resistance of the air to its motion increases. Eventually, a speed is reached where the force resulting from the rush of air matches that pulling the object down and no further acceleration takes place. Until it hits the ground, the object will continue to descend at a constant speed, called the *terminal velocity.*

If we compare bodies of the same size and shape, their terminal velocities are proportional to their weights, a result quite compatible with Aristotelian physics. A heavy steel ball falling from an airplane might require thousands of feet to achieve terminal velocity. A human body acquires it in a few hundred feet. That is the secret of sky diving, which is a long fall at terminal velocity, followed by opening a parachute to increase air resistance, and thus lower the terminal velocity to a safe value for landing.

It is far easier to study the motion of something with a low terminal velocity, such as a golf ball descending in water. There is no a priori reason why this approach to slowing down falling-body motion is any less legitimate than Galileo's inclined plane. An object can reach terminal velocity in a fraction of a second in water. Had he chosen this route, Galileo might have concluded that Aristotle was basically right after all, and decided to ignore the period of fall before the object reaches terminal velocity as a short term effect that soon goes away. Instead, he chose to ignore the equally obvious small differences in the fall of heavy objects.

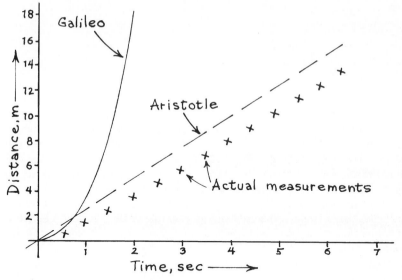

FIGURE 1-3.

Galileo made the right choice, but that is clear only in the light of the subsequent history of his science. In some respects, Aristotle's approach might have led to a better description of nature. But Galileo's idealization helped guide Newton to his law of gravity, a half century after *Two New Sciences*. Surprisingly, more than two centuries further down the line it still had the power to inspire scientific advance. In 1908, Albert Einstein found in Galileo's insight a deeper meaning than anyone had imagined.

The lesson to be learned from this exercise is that there is nothing automatic about scientific progress. Suppose this particular problem were currently on the research frontier. A team of modern Ph.D. physicists might well receive a generous research grant to study falling bodies, rapidly filling computer files with heaps of data on falling objects in all possible combinations of shape, size, weight, medium, etc. Professors and graduate students would churn out a deluge of papers for the scholarly journals, to advance their careers and justify all that expenditure of public funds.

Chances are that faced with the necessity for explaining the data to a reasonable degree of accuracy without a complete theory, they would move toward a point of view like that of Aristotle, missing Galileo's insight altogether. Science is more than a mere attempt to describe nature as accurately as possible. Frequently the real message is well hidden, and a law that gives a poor approximation to nature can prove more significant than one that works fairly well but misses an essential point.

As important as it proved to be, Galileo's idealization of falling-body motion was no more than a starting point for a new science the old man could not hope to live to see bear fruit, as he fully realized:

. . . here have been opened up to this vast and most excellent science, of which my work is only the beginning, ways and means by which other minds more acute than mine will explore its most remote corners.

Summary

Galileo wrote the *Two New Sciences* late in life, after he had been barred from writing on astronomy. The centerpiece of the book was a new, idealized mathematical description of the motion of falling bodies, a problem area for the established Aristotelian physics. In the absence of resistance all objects, regardless of weight, would fall at exactly the same speed, and gain speed through uniform acceleration, where the change in speed is proportional to the time. To make the latter statement experimentally testable, he showed that it implies that the distance fallen is proportional to the square of the time. Even then, he was obliged to test it not on freely falling objects but on a ball rolling down an inclined plane. This was an outstanding example of the distinction between experimental science and the passive observation of nature. Though Galileo's description was inadequate for understanding of the fall of real objects subject to resistance, it led to later progress by others and plays a central role in our understanding of gravity to this day.

CHAPTER 2

Toward a Science of Mechanics

If I have seen farther than others, it has been by standing on the shoulders of giants.

—Isaac Newton

Today, few of us have much trouble accepting the notion that we live on "spaceship Earth," that wonderful blue ball in the astronaut's photographs, hurtling through the cosmos at an unimaginable speed. It helps that so many of us have ample experience with smoothly moving conveyances like jumbo jets and ocean liners. If the air or sea is smooth, it is hard to tell we are moving without looking out the window.

But Galileo and his contemporaries had little opportunity for a comparable experience. Thus it took quite a feat of abstraction to realize that it is not motion itself, but deviation from smooth, steady motion, that we are able to sense. Underlying this revelation are two important physical principles. Galileo did not invent either of these principles, nor did he give them their modern names, but he did demonstrate their power in his studies of the motion of projectiles.

This was Galileo's most sophisticated contribution to science. If he was forbidden to state directly that the Earth could be moving without our being aware of its motion, he would let projectiles make the argument for him. His celebrated falling body law was actually a by-product of this research.

The two principles Galileo illustrated in his analysis of projectile motion were:

The Principle of Inertia. An object moving on a level surface will continue to move in the same direction at constant speed unless disturbed.

The Principle of Superposition. If an object is subjected to two separate influences, each producing a characteristic type of motion, it responds to each without modifying its response to the other.

Both principles are needed to counter the argument that if the Earth was moving, any time you jumped in the air you would find it impossible to come down in the same place. The Principle of Inertia allows you to continue to share the motion of the Earth, which is nearly a straight line, and the Principle

19

of Superposition ensures that your jump will respond to gravity in the same fashion as if the Earth were standing still.

The Principle of Inertia, like the description of falling-body motion, was a choice between two extreme ways of idealizing a phenomenon. The motions we observe in the real world all have some tendency to continue after whatever causes the motion is removed, but they persist for only a limited time. To cite two extreme examples, consider a stone dragged across rough ground or a hockey puck sliding on smooth ice. The Aristotelian approach to motion concerned itself with the first case and dismissed the persistence of motion as a temporary condition. By Galileo's time, however, there was considerable support, even among Aristotelians, for the idea that motion has a natural tendency to continue unless something interferes.

Both Galileo's and Aristotle's approaches have some appeal to the intuition, and there is no obvious basis for a choice between them. Once again, the most significant test in the final reckoning was not which more nearly described a larger share of the motions commonly found in nature, but which ultimately led to a deeper understanding. By accepting inertia, Galileo pointed the way to the triumphs of Newton.

One of the more dramatic illustrations of the Principle of Superposition is the observation that if a gun is fired horizontally and, at the same instant, a bullet is dropped from the height of the muzzle, both bullets will hit the ground at the same time. In the absence of air resistance, the rapid horizontal motion has no effect on the vertical motion. The winging bullet falls at exactly the same rate as the one dropped from rest, and they remain always at the same height until they reach the ground, as illustrated in Figure 2-1.

Galileo's work on projectiles began with experimental studies of exactly this case. In place of a gun, he used a miniature "ski jump," shown in Figure 2-2. A ball rolled down a ramp and flew off the end horizontally. To be sure that the ball always flew off at the same speed, he always started its roll at the same point. The whole apparatus was mounted on a support that could be raised or lowered.

Placing the ramp at various heights, Galileo measured how far away the ball landed on the floor. He found that the vertical height through which the ball fell was proportional to the *square* of the horizontal travel. To get the ball

FIGURE 2-1.

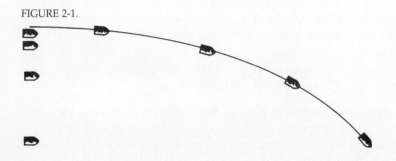

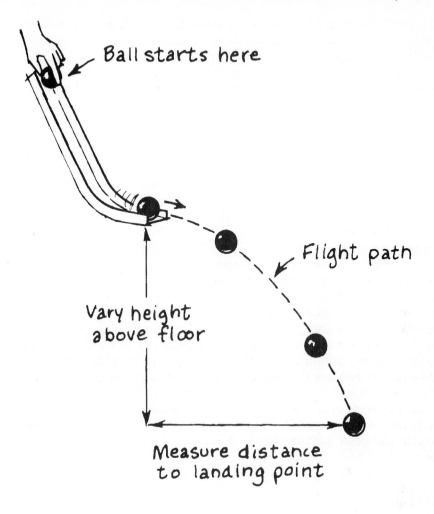

FIGURE 2-2. Galileo's apparatus for studying projectile motion.

to land twice as far away, you would have to raise the ramp not just two, but four times as high. As we shall see below, it was this observation that first led him to the rule that for falling-body motion the distance is proportional to the square of the time.

The first result of this work was that Galileo settled a long-standing argument over how to describe the path of a projectile. You probably learned in high school algebra that the graph of the formula $y = ax^2$ is a parabola. Galileo was familiar with this rule, and announced that he had demonstrated that a projectile follows a parabolic arc. For a lesser scientist, that would have been achievement enough. But Galileo had far bigger fish to fry.

Guided by the Principle of Superposition, Galileo analyzed the horizontal and vertical motions separately. The horizontal would be the simpler of the two because, by the Principle of Inertia, the ball must continue its horizontal motion at the same speed at which it left the incline. With the Sun directly above, the shadow of the ball would move at constant speed. In this kind of motion, the distance covered is simply proportional to the time. So, he reasoned, *the horizontal distance was a measure of the time of flight!*

Once he saw that, the conclusion was clear. For the vertical motion, which was that of a freely falling body, the height was proportional to the square of the time of descent. This, then, was the origin of this result, through an experimental observation. Only later did he test the rule on an inclined plane, as a further check of his reasoning. Only later still did he realize that this behavior is a consequence of uniform acceleration.

Experiment had pointed the way, but simple facts demonstrate nothing. Only with the aid of two scientific principles, neither of which were yet universally accepted, could Galileo uncover the deeper meaning of that parabolic path. *Theory without facts is blind, but facts without theory are lame.*

In most practical examples of projectile motion, an object starts its flight near the ground, with some upward vertical motion, rises to the top of its arc, and then descends. It is easy to handle this case by means of Galileo's analysis. One need merely note that in uniform acceleration, the rising projectile loses speed at the same rate the falling one gains it. It will take just as long to fall as it took to rise.

The starting point is to split the motion into horizontal and vertical components, as indicated in Figure 2-3. The vertical motion determines how long the projectile will be in flight, while the horizontal specifies how far away it will land.

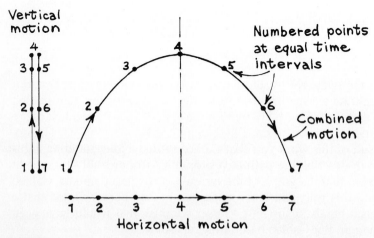

FIGURE 2-3.

For example, let a football punt leave the kicker's toe with a vertical velocity component of 20 m/s, and a horizontal one of 15 m/s. Since the acceleration due to gravity is around 10 m/s², it will take two seconds for gravity to eliminate the rising vertical motion, and another two to fall, for a "hang time" of four seconds. In that time it will travel 4 × 15 = 60 meters (about 65 yards) down the field. This is an ideal example—in actual practice air resistance will significantly shorten the punt's travel.

In general, one analyzes the vertical motion of a projectile to determine how long it will be in flight, and then uses that result to determine how far it will go. If the projectile has an initial vertical motion v_v, it will rise to the top of its trajectory in a time $t = V_v/a$. If it is moving above level ground, it returns to the same height at which it started in an equal amount of time, so the total time in flight is double this value. One then multiplies by the horizontal velocity v_h to get the distance traveled, which is usually called the "range" R of the projectile:*

$$R = 2\,\frac{v_v v_h}{a}$$

Demonstrating two new principles and uncovering the law of falling bodies were the payoff of this analysis. But to many of his readers, the discovery that the trajectory is a parabola was the selling point, because next to a circle, the parabola is the simplest curve in geometry. And in basic science, *simple* is usually taken to mean "important."

For most of the history of modern physics, simplicity has been an aesthetic criterion taken for granted. If a problem does not have a simple solution, physicists tend to dismiss it as not fundamental. Today, the advent of computers powerful enough to deal with complicated situations has begun to erode this faith in simplicity as a test of the fundamental importance of a problem.

The true path of a projectile subject to air resistance is a difficult curve with no simple mathematical formula, and is thereby dismissed as less *fundamental* than Galileo's parabola, even though the latter is a useless approximation to nature for nearly all practical purposes. Indeed, the details of projectile motion in air were not really worked out until well into the twentieth century.

DESCARTES, HUYGENS, AND MOMENTUM

The next major contribution to the understanding of motion came from the French philosopher René Descartes, who was born a generation after Galileo.

---

*For those familiar with trigonometry, this can be reduced to a statement in terms of the overall speed of the projectile v and its starting angle θ above the horizontal. Since $v_v = v \sin \theta$ and $v_h = v \cos \theta$, we have $2v_v V_h = 2v^2 \sin \theta \cos \theta = v^2 \sin(2\theta)$.

While Galileo had made a good start at building his science from the bottom up, Descartes tried to work from the top down.

Descartes fully embraced the rising rational spirit of his time, but he feared that something valuable would be lost if the majestic worldview of scholasticism was simply cast aside, with no intellectual system of comparable scope and grandeur to take its place. How were the young to be educated? It would be tragic if they concluded that reason was the implacable enemy of religious faith. Accordingly, he set himself the life's work of constructing a new general philosophy that would harmonize the new science with theology.

His methods were far more Platonist than Galileo's. He put great stress on the discovery and use of first principles by pure thought, reasoning that since God is good, He would not allow a clear thinker to have a wrong idea. As a young man of the lower gentry, raised far from the center of French culture in Paris, he bypassed a university education for training in practical matters.

For much of his life, Descartes earned his living as a military engineer, but hardly had a soldierly temperament. In his day, civil engineering was the province of the military, and his work was as concerned with roads and bridges as with guns and fortifications. Given a frail constitution, much of his life was spent in semiconvalescence, often living off the generosity of others.

To this day, French academic training encourages the use of his style of argument, which is called *Cartesian*. A basic principle is isolated and followed by impeccable deduction to conclusions of truly astonishing and often infuriating scope, a practice that does not always sit well with more empirically minded scientists in other lands.

Descartes' contributions to mathematics went far beyond those of Galileo, but his physics is best described as flawed and incomplete. The trial of Galileo took place when Descartes was in his prime, and left a strong impression. He was loath to compromise his larger goals by getting too involved in *that* controversy. Most of his physics was published posthumously, in a form that reads more like a preliminary draft than a finished work.

Indeed, much of the credit for Descartes' achievements in physics should go to Christian Huygens, the son of a Dutch diplomat at whose home Descartes was a frequent guest. Though he ultimately rejected Descartes' philosophical system, Huygens salvaged the best parts of his physics and corrected some of its more obvious defects.

The achievements of Descartes' mechanics fell far short of his projected goals, but he left two indelible marks on the history of physics. First, he directed attention to the problem of how one object transfers motion to another, an emphasis that helped set Newton on the right path. Second, in the process of studying this problem, Descartes demonstrated the power of a remarkable format for constructing a law of nature, the *conservation law.*

A conservation law might be called the scientific equivalent of the French aphorism, *plus ça change, plus c'est la même chose* (the more things change, the more they remain the same). Applied to a complicated situation in which things are constantly changing, a conservation law is an assertion that some simple quantity remains the same. Present-day physicists are so accustomed to thinking in terms of conservation laws that attempts to formulate new basic laws of physics are often phrased in this form.

A conservation law rarely provides a complete description of a process, for in spirit it implies that the details need not be considered—they will work themselves out. Herein lies its power, for it is exempt at the outset from the necessity of dealing with a phenomenon in all its complexity. To make an analogy with the social sciences, it is as if a political scientist worked out a means of predicting the exact electoral vote of a presidential candidate without being able to tell which way any particular state would go.

The seventeenth century was a wonderful era for the craftsmen who were fashioning better and more intricate machines, clocks being the most outstanding example. Like many scientists, Descartes came to regard the universe as simply a machine, the work of an Almighty Inventor. The key to understanding a machine, he concluded, was to study how motion is transferred from one part to another through direct contact. As the simplest example of this process, he focused on *collisions*. Objects are in contact for a brief moment, after which both have changed their motion.

The law that Descartes used to analyze the simple problem of the collision of two bodies was that of the *conservation of momentum*. Momentum to Descartes was the product of the weight of a moving body and its velocity. In modern physics jargon, we substitute the word *mass* for weight. *Weight* is used to describe the force that gravity exerts on an object, which varies from place to place. But the physicist's term mass is closely related to the common-language meaning of weight.

As applied to collisions, the law of momentum conservation asserts:

When two objects collide, momentum may be transferred from one to the other, but the total momentum does not change.

As an example, illustrated in Figure 2-4, imagine two objects on one of the "frictionless" surfaces beloved by writers of physics texts. One is stationary and has a mass of 3 kilograms (kg), while the other is moving at 10 m/s with mass of 2 kg. (This book will, in most instances, use the highly Cartesian SI* or "metric" system of units. We will shun English measures, a modern patchwork codification of medieval trade units, adopted to minimize commercial dislocation, a practice the French might call typically Anglo-Saxon.) Before collision, the total momentum (in kilogram-meters per second) is

$$p = 2 \text{ kg} \times 10 \text{ m/s} + 3 \text{ kg} \times 0 \text{ m/s} = 20 \text{ kg-m/s}$$

The use of p for momentum is another peculiar tradition, and the unit kilogram-meters per second unfortunately has no shorter name of its own.

The law says that after the collision this sum will be the same. That is all it says. It does not pretend to predict what speed either object will have. Further information is necessary to settle that question.

*For Système International.

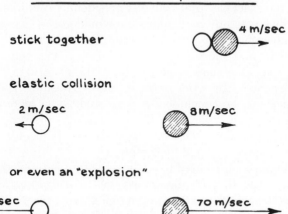

FIGURE 2-4.

The simplest case comes if the additional information is merely the qualitative assertion that the objects stick to each other. Then we have a combined object of mass 5 kg. In order to have a momentum of 20 kg-m/s, the same as before the collision, this must have a speed of

$$\frac{20 \text{ kg-m/s}}{5 \text{ kg}} = 4 \text{ m/s}$$

To give another example, the additional information could be a measurement of the speed of one object after the collision. For example, we might find that the struck one is moving at 8 m/s after the collision, in the original direction of motion of the moving object. Its momentum is then

$$3 \times 8 = 24 \text{ kg-m/s}$$

What are we to make of this curious situation? We find we actually have 4 kg-m/s more momentum than we started with! Descartes himself was baffled by this case and concluded that when a light object strikes a heavier one, it must recoil without budging it one iota, a conclusion that flies in the face of common sense. Here Huygens came to the rescue. He realized that momentum must not merely take into account the *speed* of motion but also its *direction*. Motions in opposite directions cancel one another. If we count a body moving to the right as having positive momentum, one moving to the left must have negative momentum.

Thus the 2-kg body must have momentum of −4 kg-m/s. To find its velocity, we divide the momentum by the mass, getting −4/2 = −2 m/s.

This example was not chosen at random. Note that after the collision the larger ball is moving 8 m/s to the right and the lighter one is moving 2 m/s to the left. Thus they are moving apart with a combined relative speed of 10 m/s. This is the same as the speed at which they came together, before the collision. When this happens, the collision is referred to as *elastic*. Its significance will become apparent later when we introduce yet another conservation law, that of *energy*.

This is the basis for the distinction between the words *speed* and *velocity*, mentioned in Chapter 1. When a physicist cares only how fast something is moving, *speed* is the word of choice. If the direction is specified, the correct term is *velocity*.

The law of momentum conservation would also cover a situation in which the 2-kg body is moving 95 m/s backward while its partner moves 70 m/s forward. If your intuition tells you that in this case we must be dealing with something more than mere passive balls, you are perfectly right. But that is beyond the scope of the law of momentum conservation, which sees no fundamental distinction between the three cases illustrated.

We have simplified the problem by assuming that the collision is head on and that the bodies move along the original line of motion after collision. If we consider motion in two dimensions, like the balls on a billiard table, we need simply invoke the Principle of Superposition and apply momentum conservation separately in two perpendicular directions. This complicates the mathematical description, but adds nothing to the physical principle. With the aid of the Principle of Superposition, momentum conservation may be extended to any number of objects.

The examples we have chosen all involve the motion of two bodies. But the principle can be generalized to any number of objects. Simply add up all their momenta, and at any time in the future that will be the same, as long as none of the bodies interacts with something outside the system.

THE CENTER OF MASS DOESN'T MOVE

To close this chapter, we derive a curious result of the law of momentum conservation. It is of no great significance in itself, but it provides a good example of the application of the law. Furthermore, it will prove useful later in the development of the theory of relativity:

> If the center of mass of a group of objects is stationary, no interaction among the objects can cause it to move.

The concept of center of mass was familiar to the ancient Greeks and is equally familiar to any child who has played on a seesaw, for the center of mass is nothing but another name for the "balance point." If a 90-lb girl wishes

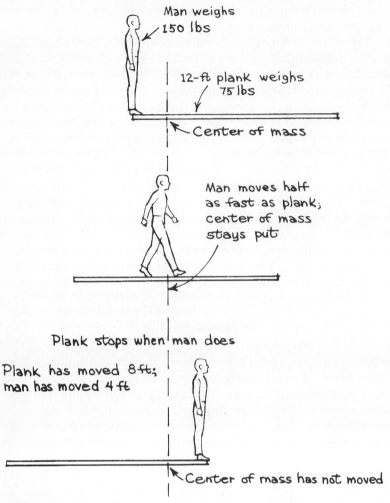

Man weighs
150 lbs

12-ft plank weighs
75 lbs

Center of mass

Man moves half
as fast as plank;
center of mass
stays put

Plank stops when man does

Plank has moved 8 ft;
man has moved 4 ft

Center of mass has not moved

FIGURE 2-5.

to balance her 45-lb brother on a seesaw, she must sit half as far from the center as he does. Stated mathematically, a seesaw is in balance if the product of weight and distance is the same for both riders.

If we consider the somewhat more difficult case of two moving objects, it is clear that if the center of mass is to stay put, the heavier must move more slowly than the lighter. If a 90-lb object and a 45-lb one are approaching each other, as long as the heavier one moves half as fast the center of mass remains at rest.

But from Huygens' point of view, this is merely the situation where the total momentum is zero. Two objects approach (or recede from) each other, the heavier at the proportionally slower speed. Their momenta remain equal and opposite. As long as no external influences come into play, the total momentum will remain zero.

As a final example, consider Figure 2-5, in which a 150-lb man is standing at one end of a 75-lb plank set on the ice so that it moves freely. As he starts to walk to the other end, the plank slips in the opposite direction. Since it weighs half as much, momentum conservation dictates it must move twice as fast. If the plank is 12 ft long, the man actually moves 4 ft while the plank slides back 8. When he gets to the other end and stops, the plank stops too. Since he has moved half as far as the plank, which weighs half as much, the center of mass has not moved.

In Chapter 11, this seemingly innocent example will serve as the basis for deriving Einstein's most celebrated formula, $E = mc^2$!

Summary

Galileo used his description of falling-body motion to analyze a more complicated situation, the motion of a *projectile*. Horizontal and vertical motions were treated separately, and Galileo's assertion that they would not interfere with one another was an early application of the *Principle of Superposition*. Horizontal motion continues at constant speed, an example of the *Principle of Inertia*. The vertical motion is that of a falling body, and the two combined lead to a parabolic path. Galileo had discovered this path experimentally and reasoned back to uniform acceleration. René Descartes and Christian Huygens developed the principle of *conservation of momentum*, which provided both a way to quantify motion and a way to analyze how motion is transferred from one object to another. One consequence of this principle is that the center of mass of a group of objects remains in the same place as motion is transferred between them. These three principles laid the groundwork for Newton's laws of motion.

CHAPTER 3

The Denouement:
Newton's Laws

He has so clearly laid open and set before our eyes the most beautiful frame of the System of the World, that if King Alphonse were now alive he would not complain for want of the graces of simplicity or of harmony in it.
—ROBERT COATES, Preface to the *Principia*

In 1665, a mere twenty-nine years after the publication of *Two New Sciences*, a student from Trinity College of Cambridge University sat in Woolsthorpe, the quiet Lincolnshire manor house where he had been born, putting the finishing touches on Galileo's "vast and most excellent science." Driven from the crowded university town by the last of the great plagues to devastate Europe, Isaac Newton made remarkable use of his period of forced isolation from academic life, returning to the small freehold where he had been born only twenty-two years before, on Christmas day.

A comparison of Galileo with Newton is a study in contrasts. Galileo's worldliness and boisterousness could hardly be more remote from Newton's haughty, mystical reserve. Galileo thought on his feet and was adept at public debate; Newton was moody and temperamental and let others fight most of his battles for him, for he had a neurotic fear of controversy and would often fly into a rage over even a minor disagreement. Galileo barely hid his skepticism behind a formal capitulation to a church tribunal without unduly burdening his conscience. Newton, on the other hand, remained throughout his life a fanatically committed Christian with a personal theology that would have branded him a heretic, had he not kept his views to himself.

Isaac Newton never knew his father, a prosperous but illiterate farmer who died two months before his son's birth. His mother's family had enjoyed higher social standing, and an uncle saw to young Isaac's education. By his midteens he had developed an irritable, withdrawn, and studious temperament. At age 17, he left school to take up management of the family estates, a task at which he failed miserably because of lack of interest. Accordingly, he was sent on to Cambridge University, where several of his relatives had been educated. Like nearly all students at the university, Isaac was expected to train for the clergy.

By the 1660s, this once-great intellectual center had deteriorated into little more than a diploma mill. Degrees were handed out to just about anyone who hung around (and paid fees) for the requisite number of terms. Most of the instructional burden fell on the college fellows, who were supposed to tutor the students. But the majority of the fellows were aspiring clergymen who had obtained their positions by patronage or out-and-out bribery, and were simply biding their time until a more lucrative church position came open. Their stipends provided a reasonably comfortable life whether they actually tutored or not, and most would have been content to remain in the college for life if they had been allowed to marry.

The new intellectual currents that were sweeping the European continent had scarcely touched Cambridge, where the now-outdated Aristotelian canon still reigned supreme. But this troubled young Isaac not a bit. Free at last from the dreary provincialism of his boyhood, he quickly dismissed the required readings, realizing that he would never be seriously examined on them. Books became his life, and he devoured with great relish the new works that defined the growing scientific revolution.

Newton's first two years at Trinity were undistinguished and uneventful. But midway through his third year he discovered mathematics through the lectures of Isaac Barrow, a swashbuckling figure who had returned from exile at the end of the period of Puritan rule. Barrow held the newly created post of Lucasian Professor of Mathematical Philosophy, the bequest of one Henry Lucas who hoped to rescue Cambridge from its intellectual slough. Within six months, studying on his own, Newton had mastered the subject to the limits of human knowledge. For the next six months he navigated masterfully through uncharted waters, until one month after the award of his B.A. degree he hit upon the central idea of what he called *fluxions*, which we know today as calculus.

Though he was reasonably adept at mathematics, Barrow's passion was to write on theology, in hopes of finding a formulation that could cool the enmity between Catholics and Protestants that had torn Europe apart. After six years, he relinquished the Lucasian chair in hopes of securing an office more to his liking. As a close friend of the executors of the Lucas estate, he was in a position to name his own successor, and he startled the serene world of Cambridge by naming the obscure, 26-year old Isaac Newton. Thus began the distinguished history of this chair, which today is held by Stephen Hawking.

For a natural recluse like Newton, the isolation of a scientist at Cambridge held no terrors. He often found the hall empty when he came to deliver his lectures, though his conscience obliged him to speak for at least a few minutes to earn his keep. Otherwise free from any onerous duties he immersed himself in thought, often laboring late into the night and sometimes going for days without bothering to eat.

He had little direct contact with other scientists, aside from an occasional dinner with a chemistry professor. Even this tenuous relationship was abruptly terminated when his guest offended Newton by telling an off-color story that involved a nun. In his heart Newton considered himself the world's greatest mathematician and philosopher, but though he yearned for recognition he dreaded the controversy and, most of all, the intrusions on his precious privacy that publication of his ideas might evoke. Unfinished manuscripts piled up in his study. For more than twenty years, the world learned of his work mainly through tantalizing fragments that appeared in his correspondence; but these alone were enough to brand him a genius.

The natural milieu for English science was London, home to the astronomer Edmund Halley, the architect Christopher Wren, and the philosopher John Locke, as well as a rival physicist, Robert Hooke. Much of their life revolved around Britain's first scientific organization, the Royal Society. Its Latin motto was *nullum in verbis,* which translates roughly as "don't take anbody's word for it." Only ideas that had met the test of experiment were to be believed.

Members of this circle had grown up in a world torn by religious strife. They had seen thousands die in battles at least nominally fought to settle the meaning of a few lines of Holy Scripture. In their eyes, traditional religion had forfeited any claim to be the ultimate fount of truth or the guarantor of a harmonious social order, and they were looking for something to replace it. The new sciences seemed to offer a better and more certain source of knowledge.

In their beliefs there was no place for the supernatural, nor any limit to the power of reason to unravel the mystery of the cosmos. Though most accepted the basic tenets of the Christian faith, they tended to be *deists,* believing in a God who had set down the laws of nature at the creation, but who did not intervene in the daily affairs of the world.* They admired the mechanical philosophy of Descartes but distrusted its adherents, whom they suspected of cloaking a new scholasticism in the garb of science in order to reassert the intellectual authority of the Church. They saw in Newton a potential champion who could meet this threat on the highest intellectual plane.

Newton had little stomach for any such struggle. By the age of 40 he had largely given up on mathematics and physics, and had spent nearly a decade secretly preoccupied with alchemy and Biblical prophecy. But a fortuitous visit from Halley jolted him out of this obsession. His scientific interest rekindled, he at last resolved to share with the world his proudest accomplishments. Eighteen months of feverish labor brought forth in 1686 the *Philosophiae Naturalis Principia Mathematica* (Mathematical Principles of Natural Philosophy), commonly referred to simply as the *Principia.*

NEWTON'S SYSTEM

The *Principia* was conceived as a showcase for Newton's solution to one key problem, the one regarded by his contemporaries as the supreme test for sci-

---

*The deist credo has been waggishly stated as "God is a retired engineer."

ence. Christopher Wren had even offered a prize for the first person to solve it. Many of the best minds in science had tried and failed.

This benchmark problem was to *explain*, rather than simply *describe*, the motions of the planets. Newton was, first and foremost, the most accomplished and creative mathematician of his age, but the task before him took more than mere computational skill. On an intuitive and computational level, he had apparently solved the heart of the mystery in his Woolsthorpe exile. But now, in his more mature vantage point, he realized that his arguments would not be fully convincing until he devised a system that could, in principle, be employed to study *any kind of motion whatsoever.* For the first time, he resolved, a work that dealt with the real world would display a logical clarity and completeness that rivaled that of Euclid's geometry.

Newton's key insight was that the work of Galileo, Descartes, and Huygens was missing one key ingredient—a way to *predict the transfer of motion.* He would rectify this omission by treating the transfer as a *continuous flow*, rather than the one-shot transaction that Descartes had considered. To name this flow of momentum between objects, he formalized the definition of a word already in wide use—*force.* In his usage, it stands for "momentum transfer per unit time."

Newton introduced his scheme, in the manner of Euclid, by formulating three postulates, which are known today as Newton's Laws of Motion. It is instructive to see these in Newton's own language, with translations into modern terminology in brackets:

> *Law I.* Every body continues in its state of rest, or of uniform motion in a right [straight] line, unless it is compelled to change that state by a force impressed on it.

> *Law II.* The change in motion [rate of change of momentum] is proportional to the motive force impressed; and is made in the direction of the right line in which that force is impressed.

> *Law III.* To every action [change of momentum] there is always opposed an equal reaction; or, the mutual actions of two bodies are always equal, and directed to contrary parts [opposite directions].

We recognize the first law as Galileo's Principle of Inertia, now elevated to a general law that applies in all directions. It sets aside motion in a straight line at constant speed as a special kind of motion, akin to rest, which needs no explanation or cause. Any deviation from this kind of motion means a force must be present. Motions of this type are now called *inertial motions.* It is now clear that Newton only came to finally accept this law during the writing of the final drafts of the *Principia*.

The third law is simply Newton's restatement of momentum conservation, in Huygens' version of the law. But the second law is pure Newton. In our era, it serves to define the SI unit of force, which has been appropriately named the *newton* (abbreviated N). Force is momentum transfer divided by time, and a newton is a transfer of one kilogram-meter per second per second.

In most transfers of momentum, the masses of both objects remain the same. Only the velocities change. In these cases, we can express the second law as

$F = ma$

In Chapter 1 we saw a car accelerating at 4 m/s^2. If that car had a mass of 1000 kg, the required force would be 4000 N.

THE CRUCIAL PEG

The real power of Newton's second law lies in the possibility of discovering laws of force that allow us to *predict* the forces that come into play when two bodies interact. This gives the scientist a clearly defined task, and a peg on which to hang the results.

For example, the force exerted by a well-made spring depends solely on how much it is compressed or stretched. If two balls have this property, their collisions will be elastic. Newton allows us to examine the brief moment of contact in detail. In the early stages the balls deform, generating a force that eliminates their relative motion. Then they spring back with a force that has exactly the same strength, restoring the original relative motion.

Once a force law is known, every detail of the motion can be predicted. Galileo had been content with an approximate, idealized description of nature, but Newton offered far more. If *all* the forces acting on an object could be taken into account, it would be possible to predict its motion exactly. One could, for example, find the law of force for air resistance. For anything more complicated than a smooth sphere, however, this turns out to be terribly difficult.

In effect, this is the ultimate reconciliation of Platonist and Aristotelian science. The observer's real world *does* exactly mirror the mathematician's ideal one, as long as all the details are understood. In practice, however, this is usually impossible. To this day, test pilots must risk their lives in order that new airplanes can be made to fly safely.

With the publication of the *Principia,* Newton's physics achieved the status of what Thomas Kuhn, a philosopher and historian of science, calls a *paradigm.* This term is now widely used (and abused!), though there is some confusion as to exactly what Kuhn intends it to mean. In this book, it will designate a group of related concepts and methods that, at least in principle, enable one to completely understand some well-defined class of phenomena.

For the Newtonian paradigm, the phenomena are all forms of motion, the concepts are Newton's laws plus the Principle of Superposition, and the method is the formulation and use of laws of force. The later chapters of this book will concern how this paradigm was replaced, in the twentieth century, by two new paradigms, relativity and quantum theory.

GALILEAN RELATIVITY

Newton's first law opened a philosophical quandary that was not resolved until the twentieth century. Because of this law it is not possible, by means of any mechanical experiment, to say absolutely whether something is standing still or moving in a straight line at constant speed.

Put another way, in Newton's physics, velocity is always *relative*, while acceleration is *absolute*. Two observers, moving with respect to one another in an inertial fashion, will obtain different values for the velocity of every object they see, but will always obtain the same value for acceleration. Because Galileo is credited as the true originator of the Principle of Inertia, this feature of Newtonian physics is called *Galilean relativity*.

A *reference frame* is a scheme for specifying the position of an object in three dimensions, starting from some point of reference. For example, the directions "north, east, and up" and their negative counterparts "south, west, and down" measured from some agreed-upon survey marker, constitute a reference frame.

An *inertial frame* is a reference frame in which Newton's first law holds true. It is to be contrasted with an *accelerated* reference frame, such as the interior of an accelerating automobile, where things will not stay put unless held in place. A formal statement of Galilean relativity can be phrased

> All inertial frames are equivalent; no mechanical experiment can tell you which is moving and which is standing still.

This disturbed Newton, and he stubbornly insisted that despite the first law, there must be *some* absolute standard for rest or motion, even if it be known only to God. This lapse in logical self-consistency disturbed philosophers from Bishop Berkeley in the eighteenth century to Ernst Mach in the twentieth. Only the work of Albert Einstein finally put this question to rest.

GOING AROUND IN CIRCLES

The ancients regarded motion in a circle at constant speed as the purest form of motion, for it was the way most heavenly bodies appear to move. Newton's scheme, however, makes circular motion a bit more complicated. It is a deviation from inertial motion, and as such requires a force.

This is intuitively reasonable. To swing a stone in a circle, you must tie it to a string, which exerts a force. The motion is perpendicular to the force, for when the string is taut the only direction the stone can move is perpendicular to the string. Airplanes bank in order to turn; the lifting force on their wings, which is perpendicular to the motion of the plane, is then no longer directly opposite gravity. It has a horizontal component that causes the plane to turn.

The only problem with this form of motion is one of terminology. Physicists refer to all deviations from inertial motion as *accelerated* motions. It is already a strain on the common-language meaning of this word to use it to encompass slowing down as well as speeding up. Now it is being stretched to cover a pure change of direction, with no change of speed! Biologists, with their love for Greek and Latin neologisms, might well have invented a new word. Keeping in mind that this is a peculiar use of a common word, let us go along with the physics usage.

It should be fairly obvious that in circular motion the force, and therefore the acceleration, is directed toward the *center* of the circle. For this reason Newton designated it as *centripetal acceleration*. The quantitative formula for the acceleration of an object moving in a circle of radius r with velocity v was discovered by Huygens and independently rediscovered by Newton:

$$a = \frac{v^2}{r}$$

To derive this formula rigorously takes a bit of calculus, and is in any event hardly worth the effort, so let us simply try to justify it intuitively. The r in the denominator shows that it takes more force to hold an object in a small circle than in a large one. Thirty miles per hour may be a placid speed on a

FIGURE 3-1.

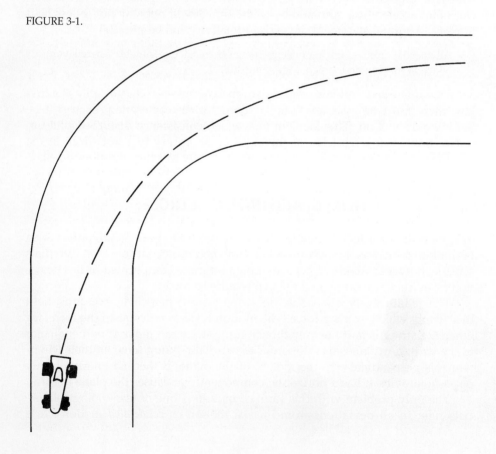

freeway interchange, but if you round a city street corner at this speed, you will hear a great squealing of tires. The v in the numerator is squared because the speed makes the situation worse in two ways: there is more velocity to change, and the change takes place more rapidly. A curve that is safe at 60 miles per hour must have a radius that is four times larger than that of one safe at 30.

As a final example of this formula, consider the path taken by a racing car rounding a curve, as shown in Figure 3-1. A good racing driver enters the turn in the outside lane, moves to the inside midway through the turn, and winds up on the outside. This path has the largest possible radius and thus permits the car to maintain the highest possible speed.

MAKING PEACE WITH GALILEO

The central concern of the *Principia* was one force law, the law of gravity. The first thing that had to be done was to accommodate Galileo's law of falling-body motion in Newton's scheme.

Constant acceleration was no problem; it simply implies a constant force, the force that physicists call the *weight* of the object. The awkward point was that the acceleration must be the same for all falling bodies. This is a surprising feature. The only way to produce it is with a force that is *proportional to the mass of the object.*

We can see this by rewriting Newton's second law as $a = F/m$. If one object is twice as massive as the other, it will also have twice the force acting on it. The two effects cancel, and the acceleration remains the same.

The acceleration due to gravity on the Earth's surface, designated by the symbol g, is about 9.8 m/s^2. It varies a bit from place to place, because the Earth is not a uniform sphere. For most purposes, 10 m/s^2 is a good enough approximation. The second law tells us that the force gravity exerts on an object of mass m is mg, so in physicists' usage, a kilogram "weighs" about 9.8 newtons.

The coincidence of a force exactly proportional to mass disturbed physicists for generations after Newton. In our own century, it disturbed Einstein so much that it led him to abandon Newtonian gravity altogether. That story will be told in Chapter 12.

THE NEWTON CULT

The publication of the *Principia* put an end to Newton's isolation, and made him a public figure. More and more he found himself drawn to London, sometimes as an official representative of his university. He found the stimulating city life more to his liking than he had imagined, now that fame allowed him to enter it on his own terms. His early support for the Glorious Revolution that dethroned King James II and brought William and Mary of Orange to the English throne entitled him to a public position, though it took several years of intrigue by powerful friends in London to secure it for him. In 1696 he abandoned Cambridge for good to take up the post of Warden of the Royal Mint. Though this was intended as a sinecure, Newton took it seriously. Finding the affairs of the mint in serious disorder, he attacked his work with his characteristic energy, displaying administrative skills surprising in a man who had spent most of his life in isolation.

Most of all Newton was on display, an intellectual adornment for a Britain reborn after decades of civil and religious strife. His London home became a salon for British and visiting foreign scientists. When Peter the Great, the tsar who was dragging a reluctant Russia into modern Europe, paid a state visit to England he made it clear that the one person he truly wished to meet was Isaac Newton. Only one vestige of Newton, the Cambridge recluse, remained: he never married, though his household was graced by the presence of a young niece whose beauty and intelligence added further luster to his reputation.

Once settled in London, Newton pretty much abandoned serious scientific work, other than completing a few of his unpublished Cambridge manuscripts. He served as president of the Royal Society, which he rescued from financial ruin by running it with a dictatorial hand. His fame protected him from rumors of religious unorthodoxy, and the true extent of his heresy, which rejected not only the doctrine of the Trinity but the very legitimacy of the established Christian churches, remained a secret up to our own century.

After his death in 1727, the Newton legend continued to grow. The cult was nurtured by the otherwise skeptical thinkers of the Enlightenment. French rationalists embraced his system even more enthusiastically than the British. The coming of the industrial revolution brought new, more practical concepts that went far beyond his own ideas, but these were easily accommodated within his intellectual legacy, the framework of the *Principia*. This monument was to endure unchallenged until the first decade of the twentieth century.

Summary

The study of motion culminated in Newton's *Principia*, which expounded his celebrated three laws. What is truly new is the second law, which defines *force* as a steady flow of momentum from one object to another. If all of the forces acting on an object can be quantified, its motion can be predicted. This elevates mechanics to what Thomas Kuhn calls a *paradigm*. The goal of physical research is thus defined as the discovery of quantitative laws of force. One by-product of this paradigm is that the meaning of acceleration is expanded to encompass change of direction as well as change of speed. Newton was helped in his career by a circle of thinkers that included Edmund Halley, Christopher Wren, and John Locke, who extolled the limitless power of human reason. In private, however, his religious and philosophical views had little in common with their own.

CHAPTER 4

The Moon and the Apple

He lives below the senseless stars and writes his meanings in them.
—THOMAS WOLFE

For most of the world's religions, the heavens are the domain of the gods. Many ancient civilizations boasted a mature astronomy before they had written language to record it. This should not surprise us, for the skies display an order and predictability unlike anything on Earth. In a confusing and uncertain world, it is reassuring to know that somewhere there is order.

Newton's fame rested on linking the divine heavens with the profane Earth, holding out the implied promise that his science might someday bring celestial certainty to Earthly affairs. To understand the nature of this achievement, let us review the history of the central problem in the *Principia*.

The earliest astronomers realized that the heavenly order was not quite perfect. The stars are well-behaved, wheeling in unison across the night sky, tracing out perfect circles as if fixed on some immense sphere. But the Sun and Moon do not keep step with them. Each slowly advances through the starry backdrop, tracing paths that repeat once in a year for the Sun and in a month for the Moon. And their progress is not quite steady. The Sun traverses half of the sky in the 184 days between the spring and fall equinoxes, while it takes only 181 to complete the other half.

The planets wander even more fitfully. Mercury and Venus swing back and forth across the Sun, accompanying its journey through the skies. Mars, Jupiter, and Saturn trace repeatable paths, but from time to time they reverse direction (go "retrograde") for a few months, as do the three remaining planets discovered in the era of the telescope. Any deviation from the mathematical purity of circular motion disturbed Plato considerably, and he enjoined the astronomers of his day to find the hidden order that would ". . . save the phenomena."

The idea that the Earth is itself a planet, and like the others circles the Sun, was one early answer to Plato's plea. It was first put in writing by Aristarchus of Samos in the fourth century B.C. The orbits of Mercury and Venus lie inside ours, so they never appear far from the Sun. The others move in orbits bigger than the Earth's and traverse them more slowly. When the Earth passes between one of these outer planets and the Sun, that planet appears to be moving backward.

But the notion that the Earth could be moving seemed to fly in the face of both common sense and the physics of the time. So when Greek astronomy had its final flowering in the work of Claudius Ptolemy of Alexandria, in the second century A.D., he proposed a scheme with the Earth at its center. The motion of each planet was compounded of several coordinated circular motions, as illustrated in Figure 4-1. Each planet moved on a circle, called an *epicycle,* the center of which moved on *another* circle, and so on to a final circle, the *deferent,* which ringed the Earth.

It cannot be emphasized too strongly that this was excellent empirical science, of the sort still practiced today under the name *Fourier analysis.* The underlying mathematics is now calculus rather than geometry, but the basic idea is the same: decompose a complex repeated motion into several simple ones.

When Nicolaus Copernicus revived Aristarchus' idea in the sixteenth century, the scheme he produced was *not* more accurate in predicting the positions of the planets than Ptolemy's. Neither was it noticeably simpler. The epicycles were still there, for planets do not move in simple circles. What made it appealing to thinkers like Galileo was that it avoided the curious coincidences of the Ptolemaic scheme, in which each planet follows an independent path, but with motions carefully coordinated to the Sun, so that Mercury and Venus stay close to it, while the other planets go retrograde only when opposite the Sun.

By Newton's time, the Copernican scheme had been refined by a new and mathematically simpler description of the planetary orbits. This was the product of two of the most unusual figures in the history of science: Tycho Brahe, who observed the planets with an unprecedented precision, and Johannes Kepler, who found in Brahe's observations a new response to Plato's plea.

THE MANGY DOG AND THE MAN
WITH THE GOLDEN NOSE

Johannes Kepler was born eight years after Galileo in the town of Weil der Stadt in the Duchy of Wurttemberg, in southwest Germany. His father was a ne'er-do-well soldier of fortune who somehow won the hand of the daughter of the town's mayor. He simply vanished when Johannes was in his teens.

The Dukes of Wurttemberg were enlightened rulers, and provided scholarships to the University of Tubingen for bright students who, like Kepler, could not otherwise afford to go. There he fell under the influence of the astronomer Michael Maestlin, who spotted and cultivated Kepler's mathematical talents.

Kepler nonetheless persevered in his intention to study for the ministry, until it became clear that he could not fully accept the Lutheran credo. To

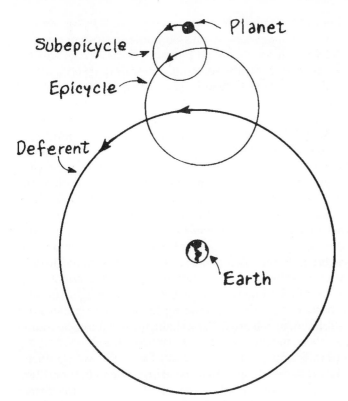

FIGURE 4-1. Ptolemy's depiction of motion of one planet.

spare everyone embarrassment, the university senate posted him as mathematics teacher to the Lutheran school at Graz in western Austria. He also served as court mathematician to the Catholic rulers of the province. A few lucky astrological predictions kept him in their good graces, while a book of mathematical speculations about astronomy, the *Mysterium Cosmographicum*, earned him a modest reputation among astronomers.

Though rewarded by fame, Kepler was never to know either prosperity or peace of mind. Tortured by real and imaginary illnesses, denied his salary by some of his patrons, forced to break off his labors to defend his somewhat dotty mother from a charge of witchcraft, his life was a frantic juggling act to keep the wolf from the door and the demons in his mind at bay. He chose for his own person the metaphor of the *mangy dog*.

The turning point in Kepler's life came in 1600 when a ban on Protestants forced him to flee Graz for Prague, where he was offered refuge as an assistant to Tycho Brahe, the foremost astronomer of his time. Tycho had deserted his native Denmark for rather different reasons.

Tycho's origins were as splendid as Kepler's were mean. Born of the union of two of the leading families of the Danish nobility, he was pointed by family tradition toward a career in the service of the crown. But as a 14-year-old student at the University of Copenhagen, he witnessed a total eclipse of the Sun, always an awesome and terrifying sight. What most impressed Tycho was the fact that it had been predicted, with seemingly uncanny precision, by astronomers. Tycho decided that any calling capable of such a feat was well worth devoting one's life to.

Studying astronomy on the sly while supposedly preparing for a legal career, Tycho soon discovered that the precision that had attracted him to the subject still left a great deal to be desired; in particular, tables of planetary motion could be trusted only for a few decades or so. After this, they were likely to be off by days or even weeks. He correctly perceived that any improvement must rest on better instruments, and he combed northern Europe in search of artisans who could build them.

His big break came in 1572, in his twenty-fifth year, when a supernova erupted in the northern sky. It was a sensational event, a new star that for months far outshone every star and planet in the heavens. His superior measurements proved that this astounding object lay far beyond the atmosphere or even the planets, in the domain of the supposedly unchanging stars. This discovery became the basis for a book expounding his scientific philosophy, which made him an international celebrity. Capitalizing on his fame, he toured Europe's great centers of learning, and resolved to settle in Basel.

Compared to southern Europe, sixteenth-century Denmark was an uncivilized nation, but it was a rich one. Its prosperous farms fed much of northern Europe. The Protestant Reformation had placed much of its wealth, the former church lands, in the hands of King Frederick II. This sovereign was determined not to lose his most celebrated subject, and made Tycho an offer unprecedented in the history of science. He would get the island of Hven as a site for his observatory, the income from several parishes to support it, and generous grants from the royal purse to build the finest instruments skilled hands could fashion. Tycho adorned the island with a splendid Italian Renaissance palace, in sight of Hamlet's brooding medieval castle of Elsinore.

This was "big science," even by present-day standards. Tycho presided over a large staff of artisans and students, with duplicate equipment that permitted four simultaneous independent observations, all but eliminating human error. Tycho and his students improved the precision of astronomy, frozen at ten minutes of arc since the days of Ptolemy, to one minute. All this was done with the naked eye, for Galileo's telescope was still several decades in the future.

Tycho was far from a hero to many of his noble peers. A great bear of a man, fully endowed with the arrogance of his class, he upbraided the nobility for their preoccupation with hunting, gluttony, lechery, and dueling (Tycho himself, in his youth, lost most of his nose in a duel, replacing it with one fashioned from an alloy of silver and gold). He further scandalized them by marrying the daughter of a peasant from one of the Brahe estates, which by Danish custom made his children illegitimate.

For more than twenty years Tycho scanned the skies from Hven. The end product was a superb catalog of the thousand brightest stars, and the most accurate and continuous log of planetary positions ever taken. But King Frederick passed on and his son Christian IV had little patience with his father's expensive hobby. In addition, he was strapped for revenue for a war with Sweden that was going badly.

Citing as an excuse Tycho's exactions of labor from the peasants of Hven, which were undeniably excessive, and his neglect of the churches given over to his care, Christian deprived the observatory of much of its income. Incensed, Tycho left for Prague to enter the service of Rudolf II, the Holy Roman Emperor, who was himself an amateur astronomer.

Tycho carried with him his instruments and his precious tables of observations. He hoped to crown his fame with his own cosmic model, which would incorporate the virtues of the Copernican scheme while allowing the Earth to sit still, as Aristotelian physics said it must. Kepler was hired to carry out the arduous calculations required to complete this task. Confident of his own powers of persuasion, Tycho cared little that his new assistant was a convinced Copernican.

Tycho's model was simplicity itself. While the Sun would continue to circle the Earth, as in Ptolemy's scheme, all the planets would move in orbits centered on the Sun. From the observational point of view, this was indistinguishable from the Copernican scheme—the apparent position in the sky of the Sun and planets would be the same. A mechanical model of Tycho's system would be the same as the Copernican. The only difference would be whether the Sun or the Earth was attached to the support stand.

This scheme was widely accepted among astronomers, especially by a group of Jesuits who had used the strong discipline of their quasi-military society to coordinate a well-organized program of observation. It was on the basis of Tycho's scheme that they rejected Galileo's arguments for the Copernican system in the *Siderius Nuncius*. Galileo's telescope had revealed that Venus shows phases like the Moon, with the illuminated side always facing the Sun. But that merely proved that Venus moved in an orbit centered on the Sun. It was still possible to believe the Sun orbited around the Earth.

Only physics, not observation, could settle this question, and Kepler was one of the first astronomers to fully grasp this. In his most important work, *Astronomia Nova* (New Astronomy), he proclaimed "You physicists, prick up your ears! I am about to invade your territory."

KEPLER CHARTS THE MOTIONS OF THE PLANETS

Tycho was to live less than one year after Kepler's arrival in Prague. Kepler inherited both Tycho's job and his notebooks, though his rights to the latter were challenged by Tycho's heirs. What he did with them was a tour de force of data analysis, one that looks impressive even from the vantage point of the computer age. His sensitivity to both the value and the limitations of precision measurement was centuries ahead of its time.

The record of Kepler's labors survives in two great works, *Astronomia Nova* and *Harmonice Mundi,* and in notebooks that detail 900 pages of laborious hand calculations. They are a unique legacy, for Kepler carefully reported not just his conclusions but a complete account of the tortuous path he took to arrive at them, replete with false starts, blind alleys, and wrong hypotheses discarded only after months of patient toil. Interspersed are fragments of verse in which he castigates himself mercilessly for his temporary failures and exults wildly in his final triumphs.

Most of all, however, Kepler's writings reveal the mystical vision that drove him to this great rational triumph. He put the Sun at the center of his universe because, as the giver of light and life, it was closer to God than the base Earth and thus more worthy of the honor. He pursued the planetary orbits relentlessly, certain that they would provide a divine lesson in geometry and the laws of musical harmony. This, the dominant vision of Kepler's life, proved a false lead. But on the way to this personal disappointment, he left behind a description of planetary motion in three "laws" that endure to this day. Since the first two are geometric, they are illustrated in Figures 4-2 and 4-3.

1. The planets travel in ellipses with the Sun at one focus.
2. The area swept out by a line drawn from the Sun to a planet is the same in equal time intervals.
3. The square of the length of each planet's year is proportional to the cube of the major [long] axis of its orbit.

The second law indicates how the planets vary in speed through their orbits, moving fastest when nearest the Sun and slowest when farther from it. The third shows that the outer planets move more slowly in their orbits than the inner ones, and thus the length of the year increases more rapidly than the size of the orbit. If there were a planet with an orbit four times the size of Earth's, its period would be eight years, since $4^3 = 8^2 = 64$.

Kepler's third law also applies to satellites of the Earth. The space shuttle, launched into low orbits only a bit larger than one Earth radius, takes about an hour and a half to go around once. At 6.8 Earth radii, the period is 24 hours; a satellite placed in this orbit above the equator will appear to be stationary in the sky. That's why this orbit is used by communications satellites: a dish antenna can be aimed once and will continue to track the satellite. Finally the Moon, in an orbit sixty times larger than the radius of the Earth, takes twenty-nine days to complete one month. In each case, if you take the ratio of orbit radii and cube it, you get the square of the ratio of orbital periods.

Though Kepler's orbits are ellipses, they are not very elongated ones. In a scale drawing, it would be hard to distinguish them from circles by the naked eye. But the small deviations from perfect circles are important, because they allowed Newton to convincingly demonstrate that only one kind of force could produce such motions.

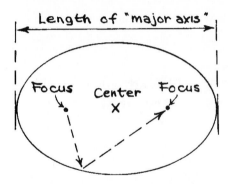

FIGURE 4-2. Structure of an eclipse. If it were lined with a mirror, any light ray from one focus would reach the other; hence the term focus.

Kepler's abandonment of uniform circular motion was a daring break with the traditions of astronomy. It was guided by physical insight: he could not stomach epicycles, because he wanted the planets to be moved and steered by forces emanating from material objects. The center of an epicycle, whether it be in Ptolemy's, Copernicus', or Tycho's scheme, is nothing but a moving geometric point. Kepler even insisted that gravity was a universal force of attraction among all objects, our modern view. But he had not yet fully embraced inertia, so he also imagined a propulsive force to keep the planets moving.

Today we recognize Kepler's second law as an example of the conservation of *angular momentum*, a measure of rotary motion.* Dancers and figure skaters put this law to good use when they pull in their arms close to their bodies in order to increase the speed of a spin. For a planet, it means the planet moves fastest when it is closest to the Sun.

Kepler's laws were eagerly embraced by the new philosophers of the seventeenth century, except for Galileo, who stayed true to his Platonism by embracing the empirically indefensible circular orbits in his *Dialogue*. Perhaps he was jealous of Kepler's achievement, which threatened his status as the foremost advocate of the Copernican view. The followers of Descartes claimed that the laws could be explained by assuming that "empty" space is filled with a material substance moving as a vortex (whirlpool) centered on the Sun. But they never succeeded in constructing a mathematical proof to back up this claim. By Newton's time, Kepler's laws were recognized as the supreme achievement of empirical science, and explaining them was accepted as the supreme test that any new philosophy must meet. That was the basis for Wren's offer of a prize for the solution of this problem.

---

*The formula for angular momentum is $mvr \cos \theta$, where r is the distance to the center of rotation and θ is the angle between the direction of motion and the line drawn to the center.

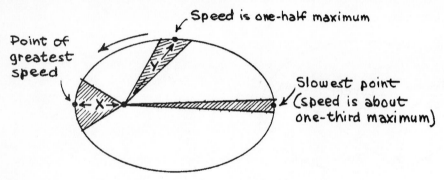

FIGURE 4-3. Areas swept out in equal times are equal.

While he may not have gotten the physics quite right, Kepler had set the table. It was Newton who would fill it with a sumptuous feast.

THE LAW OF GRAVITY

In Chapter 3, we noted how Galileo's law forced Newton to make gravity proportional to mass. Accepting Kepler's notion of a universal attraction, it seemed logical to make it proportional to the masses of *both* objects. This was a blessing in disguise, for Newton had no way of estimating the mass of the Earth, Sun, or planets. With Galileo's law extended to heavenly bodies, Newton could ignore the planetary masses and move on to Kepler's laws.

A major portion of the *Principia* is devoted to detailed, difficult geometric arguments designed to explain the significance of Kepler's work. First, he proved that the second law requires that the planets move subject to a force *directed toward the Sun.*

Newton then demonstrated that Kepler's first and third laws were possible only for a force that diminished with distance in one particular mathematical way. It must wane in proportion to the *square of the distance.* Move twice as far from the Sun, and the force would be four times weaker. This is known as the *inverse-square law.*

Newton's law of gravity can be summarized in one formula:

$$F = G \ \frac{mM}{r^2}$$

where F is the force of attraction between two objects, m and M their masses, r the distance between them, and G is a fundamental constant that sets the strength of the force.

Wren and Halley had guessed the inverse-square character of the law of gravity. First, if you treat the planetary orbits as circles, ignoring the slight deviations, a little simple algebra will get you from an inverse square law to Kepler's third law. Second, an inverse square law seemed reasonable. Anything (such as light) that radiates equally in all directions will spread over an area that increases as the square of the distance, so it must diminish in strength in the same fashion.

To show that the inverse-square law gave all three of Kepler's laws took all the mathematical skill that Newton could muster. For example, he had to prove that for an inverse-square law it is the distance between the *centers* of spherical objects that counts. These proofs would have been a considerable achievement, admired by Newton's fellow mathematicians, but unlikely to have a strong emotional impact on the public at large. What crowned Newton with fame was finding a connection between this force and the familiar pull of gravity we experience here on Earth.

The perceptive reader may have noticed by now that while the Moon is featured in the title of this chapter, it is the one heavenly body whose motion has not yet been discussed. All the cosmic schemes we have outlined concede that this one body, at least, orbits the Earth. By Newton's time, its distance from Earth was known to reasonable accuracy. This gives the Moon a crucial role in our story, for it provided Newton his indispensable link between the heavens and Earth.

THE MOON IS THE KEY

In his later years, Newton always maintained that he had happened upon the law of gravity at the age of 23, when the fall of an apple at his Woolsthorpe home provoked him to wonder whether the force that made it fall might extend as far as the Moon. He knew that the distance to the Moon was sixty times the radius of the Earth. A quick calculation to check this idea (in English units, of course, though we shall stick with SI) showed him the Moon's acceleration was far less than that of the apple. The Moon is 380 million meters from the center of the Earth, moving at an average speed of 1016 m/s. By the formula for centripetal acceleration,

$$a = \frac{v^2}{r} = \frac{(1016 \text{ m/s})^2}{380,000,000 \text{ m}} = 0.00272 \text{ m/s}^2$$

much smaller than the typical value of 9.8 m/s^2 at the Earth's surface. The naturalness of an inverse-square law then popped into his mind. Since the apple is just one Earth radius from the Earth's center, while the Moon is sixty, the acceleration of the moon should be $60^2 = 3600$ times smaller:

$$a = \frac{9.8 \text{ m/s}^2}{3600} = 0.00271 \text{ m/s}^2$$

Given the small variations of all the numbers involved, this was close enough. The result was electrifying. A quantitative link had been found between a phenomenon on Earth and one in the heavens. The Moon was the one heavenly body known to orbit the Earth, and it had been shown to move in response to a force that could be connected quantitatively to the gravity that caused the apple to fall. A new and truly universal science had been born. Why, then, did Newton wait another twenty years to tell the world?

His own explanation makes for an odd tale, resting on a peculiarity of English measures. Away from the libraries of Cambridge, Newton had to fall back on his prodigious memory. He knew the acceleration due to gravity in feet per second squared. To compare this to the Moon's acceleration, Newton needed to know the distance to the Moon in feet, and its speed in feet per second. He got this from two other facts: a mile has 5280 feet, and the circumference of the Earth is 21,600 miles. But there are *two kinds* of miles: 5280 feet is a *statute* mile, whereas 21,600 is the Earth's circumference in *nautical* miles, which are 6071.1 feet each! He was left with a 15 percent discrepancy, which caused him to drop the problem until solicited by Halley to try for Wren's prize.

This story is a bit hard to swallow—not that he could have made such an error, but that it would have caused him to give up. It must be remembered that this was a tale told by the later Newton, the London social climber. It supports the claim that Newton discovered the inverse-square law long before Halley and Wren guessed at it.

Be that as it may, this peculiar tale in no way diminishes the luster of Newton's accomplishments. Today, the study of Newton's private papers has revealed that while he owed many of his key insights to his hiatus at Woolsthorpe, it took some time to fully realize their consequences. Still, prior to Newton there had never been a work quite like the *Principia*, and there have been few comparable scientific achievements since. And we must not forget the calculus, the mathematics of continuous change, the first body of mathematics that was wholly a product of European culture. Newton worked out most of his problems with the aid of this tool, but did not actually use it in the *Principia*, which he wrote for the broadest possible audience.

HYPOTHESES NON FINGO

If the theory of gravity was Newton's greatest accomplishment, one feature of it was to provide his most severe trial. This was the notion of *action at a distance*, that two bodies could attract one other with nothing intervening but empty space. Privately, Newton dismissed this idea as absurd and was sure that some "agency" must transmit the force, though he had no clue to what it might be.

But his public stance was far different. With his disdain for the give-and-take of scientific debate, he opted out of the dispute with a haughty "*hypotheses non fingo*," by which he meant "I do not engage in idle speculation."

In his own time, this disclaimer left Newton vulnerable to attack from the Cartesians, who scented in it the very kind of "occult" explanation of nature that science was supposed to have banished forever. Forces, they felt, should be transmitted by direct contact. Newton regarded such literal mechanical models of the universe as naive, and was willing to accept unseen influences that fell outside their scope.

Indeed a dominant theme in the history of modern physics has been the struggle to expunge the intangible and unseen in favor of things substantial and material. On several occasions, this effort has seemed on the verge of success, only to be thwarted once again. Today, as we shall see in the final chapters of this book, the insubstantial has triumphed utterly, and the "agency" that Newton so cavalierly dismissed has taken primacy over matter itself.

A TRULY UNIVERSAL FORCE?

Newton's law of gravity is an outstanding example of a scientific law that is far more than a simple summary of observed facts. It contains elements that were put in for formal or aesthetic reasons. Many things it implies were untestable by experiment in his day, and some remain so today.

In all the cases Newton dealt with, one or both of the pair of attracting objects was of astronomical size and unknown mass, for gravity is in fact a very weak force. The attraction between objects small enough to pick up and weigh is undetectable by any but the most sensitive instruments. Thus he had no way of obtaining a numerical value for the constant G. The assumption that gravity is a universal force in which both partners are equivalent was an *aesthetic* choice.

It was not until early in the nineteenth century that Henry Cavendish developed an instrument that could detect the feeble force between two objects in the laboratory. By then the faith in Newton's law of gravitation was so great that Cavendish did not call his experiment a "confirmation of the law of gravity," but instead styled it "weighing the Earth." Since he had measured the force between objects of known mass a known distance apart, he could solve for G as the one remaining unknown in the equation. Once G was known, the force on an object could be used to calculate the mass of the Earth.

Indeed, to this day the only way we know the mass of any astronomical body is by assuming that the law of gravity is correct. If the dependence on mass is somewhat different for very large objects, we would have no way of knowing it. Even the inverse-square character is not established beyond challenge: in the 1980s some measurements of variations of the Earth's gravity over distances of hundreds of meters suggested that for short distances the law might be wrong by as much as 1 percent. These measurements have been disputed, but such is our faith in gravity that their authors never claimed to have found a violation of Newton's law, but simply to have discovered a *new* force that interferes with it.

Table 4-1 breaks Newton's law of gravity into seven statements with the justification for each. Empirical evidence is cited in **boldface type,** to emphasize how much the law rests on other considerations.

TABLE 4-1.

There is a force	Required assumption to fit gravity into Newton's system. Whenever motion is accelerated, there must be a force. There is no other basis for this assertion.
of attraction	Newton proved that **Kepler's second law** shows the force points along the line joining the Sun and planet.
between all objects	130 years would elapse before the force between two objects one can pick up and handle was observed. But assuming no distinction between a planet and a rock other than mass was aesthetically pleasing.
across empty space,	Most objectionable aspect of the theory to Newton's contemporaries. Regarded by some as a return to prescientific "occult influences." Lacking any observable mechanism to transmit the force, Newton found it unavoidable.
proportional to m	In cases available to Newton, only motion of the smaller object was observable. The evidence that the force on it was proportional to mass was essentially negative: **Galileo's falling-body** law shows no dependence of acceleration on mass.
and to M	There was no way to measure the mass of the Sun or a planet, so the dependence on the larger mass was untestable. It remains largely so today. But treating both objects symmetrically preserved an aesthetic quality appropriate to a universal force.
and to $1/r^2$.	**Kepler's first and third laws** both indicate that the accelerations of planets vary in this fashion. The **comparison of the Moon's acceleration to that of falling bodies on Earth** shows that Earth's gravity also obeys the rule. Thus this feature had strong empirical support.

Summary

The conflict between an Earth-centered and a Sun-centered universe was brought to a head through the work of Tycho Brahe and Johannes Kepler. Tycho improved the precision of astronomy through better instruments and a team approach. He hoped to establish a system in which the Earth was at rest, but all the other planets moved in orbits centered on the Sun. Observationally, such a system would be indistinguishable from one with the Sun at its center. After Tycho's death Kepler used his data to find the correct mathematical form of the orbits in a Sun-centered system. The central problem of Newton's *Principia* was to derive these orbits from his law of gravity, in which force was proportional to the masses of both objects and the inverse of the square of the distance between them. Through the Moon's motion he was able to link this force to the acceleration of gravity on Earth, the first quantitative connection between celestial and terrestrial phenomena.

CHAPTER 5

The Romance of Energy

may God us keep
from single vision, and Newton's sleep.
                    —WILLIAM BLAKE

Throughout its history, science has experienced a creative tension between two contrasting approaches to nature, which today go by the names *reductionism* and *holism*. Reductionism holds that true understanding of nature is to be sought beneath the surface, in the hidden workings of a few simple principles. Holism takes nature as it is, and looks for the complex web of interconnections between its parts, connections that it holds to be more important in shaping the world we live in than the parts themselves.

Put another way, reductionists view the world as a *machine,* and a machine is best understood by discerning the operations of its parts. Holists see it as an *organism* whose parts draw significance from the way in which they relate to the functioning of the whole.

Since at least the time of Galileo, physics has been unabashedly reductionist in its approach. Still, from time to time holistic ideas have had a significant impact on its development, though in these instances it eventually returned to its reductionist roots. The story of *energy* is an outstanding example of this process.

The century following the publication of the *Principia* was one of unbroken triumphs for Newtonian physics and the astronomy it served so well. Calculus was honed into a fine analytic tool and the motions of the Moon and planets were charted to astonishing precision. Nonetheless, toward the end of this period, quite a few scientists began to realize that Newton's laws and the concept of momentum were not as complete a science of motion as one might want. The Newtonian paradigm had to be extended to deal with a number of practical questions.

The problem was that Newtonian physics simply took force as a given and went on from there. It offered no way to even ask a question that the pioneers of the industrial revolution confronted daily: What does it take to generate a force? Engineers and inventors wanted to move machines, goods, and people more effectively, and all Newton had to offer them was the assurance that however they managed to do it, an equal and opposite motion would inevitably arise in the process. In most instances, this was of little practical help.

At the same time many poets and philosophers, who once applauded the liberating influence of the Newtonian spirit, began to speak of its darker side. The cool analytic method, seeking precision through a reductionist process of dissection, often lost sight of the beauty and unity of nature. While they still extolled the power of human reason, they feared a sterile rationalism that seemed to leave no room for the emotional wellsprings of creative thought.

The impact on physics of these two criticisms, which arose from seemingly opposite poles, would be a happy fusion in the concept of energy. Today, energy ranks as the central unifying core of physical science, one by which physics reaches out to embrace all other sciences and the practical world.

MUST WE "PAY" FOR A FORCE?

A few examples will serve to reveal the practical shortcomings of strictly Newtonian physics and point the way to new concepts that make it more practical.

Consider first a bullet fired from a gun. Momentum conservation requires that the gun recoil with momentum equal and opposite to that of the bullet. Their combined momentum was zero before and remains zero afterward. Yet simple common sense tells us that something significant has changed. Something was taken from a bit of gunpowder and was transformed into the motion of the bullet and gun. In the process, the powder was transformed and lost the capacity to do this again. Somehow, physics should be able to make a clearer distinction between the situation before and after.

Another very familiar practical use for a force is propelling a car. Much of the time, the car moves at a fairly constant speed, yet some force is still needed to overcome friction and air resistance. From the point of view of Newton's laws, this is a thoroughly uninteresting case. The motive force exactly balances the resistance, so the net force and acceleration are zero. The only recourse for the designer is to measure the force required and see to it that the motor is up to the task.

Newton's laws are not entirely useless in this situation. They do tell the designer how much additional force is needed to accelerate the car and how much force the brakes must exert to stop it. They also give a reminder that these forces produce equal and opposite reactions on the road.

But we know that the motor needs fuel while the brakes do not. The brakes, however, do heat up, and to get rid of this heat air must flow over the brakes. The questions of how much fuel the motor must burn and how much heat the brakes must shed seemingly lie beyond the whole science of motion.

When the car is rounding a curve at constant speed, the situation is entirely different. The driver hardly exerts any effort at all, even in a car without power steering. But if the curve is taken at close to the maximum safe speed, the acceleration is nearly as great as it is in a jackrabbit start with the engine roaring.

To speed up a car, we take something from the fuel; to stop it, we discard heat; to simply change direction, nothing is taken from or lost to the outside world. Yet in each case, the force has about the same *strength*. It is the *direction* of the force that makes these situations different. A forward force must be paid for; one to the rear obliges us to get rid of something; while a force perpendicular to the motion costs nothing.

Using these examples as a guide, we can outline a three-step program for a more practical science of motion:

1. We need a new conservation law based on a nondirectional measure of motion, which, unlike momentum, does not cancel out motions in opposite directions.
2. The connection of this measure to Newtonian physics must take into account the direction of the force relative to the motion, so that a forward force has a positive effect, a backward force a negative one, and a perpendicular force no effect at all.
3. Finally, we must find connections to things that seem at first glance to have nothing to do with motion, such as heat or the power that resides in fuels and explosives.

It is time to give a name to the mysterious something we are looking for, which manifests itself sometimes as motion and sometimes in other forms. It is called *energy,* a word borrowed from the same poets and philosophers who assailed Newtonian science. We will begin our quest for energy by building outward from the Newtonian paradigm.

WORK AND KINETIC ENERGY

The first two steps are taken care of by expanding the Newtonian language of motion. We introduce a new measure of motion that stands on an equal footing with momentum but does not replace it. It is called *kinetic energy* (*K*), which is defined by the formula:

$$K = \frac{1}{2}mv^2$$

Squaring the velocity is what makes kinetic energy a nondirectional quantity. Whether the velocity is positive or negative, its square is always positive. The kinetic energy of motions in opposite directions does not cancel out.

To see that this is also an intuitively satisfying way to measure motion, consider the case of a recoiling gun. Newton tells us that the bullet and gun get equal and opposite momenta: if the gun is 100 times heavier, the bullet moves 100 times faster. But common sense cries out that somehow, that bullet packs more wallop. We don't mind absorbing the recoil of the gun, but we certainly don't want be hit by the bullet. Since kinetic energy depends on the *square* of the velocity, the gun and bullet are not equal by this measure. The bullet gets 100 times as much kinetic energy as the gun, and absorbing this energy is what damages the target.

The next step is to define a quantity called *work*, which is a measure of *energy transfer* by the action of a force. Work is calculated by multiplying the force by the distance an object moves *in the direction the force acts.* For example, if the force is gravity, we count only the vertical distance moved. This definition is most easily put in mathematical form by using the cosine function from trigonometry:

$$W = Fx \cos \theta$$

where θ is the angle between the force and the direction of motion, and x the distance moved. If θ is in the forward region, between 0 and 90 degrees, the cosine is positive and the work adds energy to the object being pushed. From 90 to 180 degrees the cosine is negative, and energy is removed. At exactly 90 degrees the cosine is zero, so no work is done. This is precisely what was demanded in step 2 of our three-step program.

For example, if a car accelerates from rest, the work transfers energy to the car. The force is in the direction the car is moving, so $\cos \theta = 1$ and the work is simply Fx. If there were no other force acting, we could simply equate this to kinetic energy, but in the spirit of this more practical approach to motion, it must be pointed out that not all work appears as kinetic energy; some is always lost to friction and air resistance.

The concept of work is equally applicable to the Aristotelian scenario of a car moving at constant speed. In this case, the force generated by the motor is still doing work, but this work does not go into increased kinetic energy. The work, however, does not simply disappear; it is transferred to other forms of energy.

WATTS, POWER, AND ARISTOTLE

Because energy is a commodity in our civilization, and a terribly vital one at that, it is worthwhile to pause briefly to take up the practical question of energy units.

The metric unit of energy is the *joule* (J). It is defined as the work done by a force of 1 newton operating over a distance of 1 meter. The joule, however, is a unit used mainly by scientists, because it is far too small for commercial use. The most common energy unit in commerce is based on the metric unit for something called *power.*

Power is defined as the *rate* at which energy is transferred. Thus it plays the same role in our new extended paradigm that *force* played in the original Newtonian one. There is an important difference, however. In Newton's scheme an unopposed force always leads to a transfer of *motion* from one object to another, while power can mean other things, such as the transformation of electrical energy into light.

If 1 joule of energy is transferred per second, the power is one *watt* (W), named after James Watt, the designer of the first really practical steam engine, which transferred heat energy into motion. If you multiply watts by a unit of time, you get a unit of energy. To meter electrical energy, it is common to use the hour as the time unit. But even a watt-hour (Wh) is a terribly small amount of energy, so the unit in commercial use is the kilowatt-hour (kWh), 1000 times larger. Since there are 3600 seconds in an hour, a kilowatt-hour is 3.6 *million* joules! Since this much energy is delivered to your home for only a few cents, you can see what a small unit the joule is.

The *horsepower* (hp) is another familiar unit of power. It is not an official SI unit, but is usually defined as 750 W. The horsepower rating of a car's engine represents the maximum power it can deliver. Unless you have a stick shift and are an unusually aggressive driver, you have never run the engine of a car at anywhere near its rating.

A close examination of the concept of power reveals one of the things that we have appended to Newtonian physics. If we consider motion at constant speed, we will find we have resurrected the shade of our old friend Aristotle! When a car moves at constant speed on a level road, work is being done at a constant rate, and the power required is the force times the velocity:

$$P = \frac{Fx}{t} = Fv$$

For example, a typical passenger car moving at 20 m/s (45 mph) on a level surface requires a force of 900 newtons (about 200 pounds) to keep it moving. The power required is $P = Fv = 900 \times 20 = 18,000$ W $= 24$ horsepower. Since that is considerably less than the horsepower rating of a car of this sort, the engine is not laboring heavily under these conditions.

The force impelling the car is exactly equal to the force of resistance opposing it. Making this substitution, and solving for the velocity, we are back to a familiar relationship: *velocity is the ratio of power to resistance!* It is *power*, rather than Newtonian *force*, that should be used to quantify propulsive effort in Aristotle's physics. There was some value, after all, in his approach to the study of motion. Still, we must not forget that this formula applies only to the particular case of an object moving at constant speed with propulsive force balanced by resistance. It cannot pretend to the generality of Newton's laws.

THE MANY FACES OF ENERGY

The third step in the development of the broader energy concept begins with the observation that when motion disappears through friction, heat is produced. This led Robert Boyle, a generation before Newton, to conjecture that heat might itself be the motion of the invisible atoms that make up matter. But the eighteenth-century scientists who put the study of heat on a quantitative basis found it intuitively satisfying to think of it as a substance called *caloric*, and this view generally prevailed well into the nineteenth century.

What changed the minds of most scientists was the work of James Joule, a Scot who originally accepted the caloric theory. In 1842, Joule demonstrated that a paddle wheel turning in a sealed, insulated tank could produce heat indefinitely, and the amount of heat generated was strictly proportional to the amount of work done in turning the paddle wheel. Since heating water was the basis for the accepted unit of heat, the *calorie*, Joule established a quantitative connection between heat and motion. That is why his name is used for the SI unit of energy.

Joule's discovery came while the industrial revolution was in full swing. The symbol of the age was the steam engine. The practical demands of technology provided a strong motive for a theory of this process, but in the early days the technology was far ahead of the science. Engineers made improvements in steam engines, and the scientists came along later to explain why they worked. The relation of science to technology is many-faceted, and the view that science is the engine that drives all technological advance is a bit naive. The two advance hand in hand, and communications between scientists and technologists must be a two-way street.

Energy exists in, and can be transferred to, many other forms, for example, electrical and chemical. Thus, a complete law of energy conservation takes a holistic form that quantitatively enshrines the connections between different phenomena:

$$[\text{motion}] + [\text{heat}] + [\text{electricity}] + [\text{sound}] + [\cdot \ \cdot \ \cdot] + [\cdot \ \cdot \ \cdot] = \text{const.}$$

For each form of energy, the procedure for finding a quantitative measure is different. For heat, it involves the temperature, the amount of material, and a number called the *heat capacity*. For electricity, it involves the voltage and the current. The formulas vary, but ultimately all can be measured in the same units. To a reductionist, the fact that all these disparate elements of nature, each with its own quantitative measures, add up in this fashion is almost too good to be true: it is an invitation to look to a deeper, hidden level of reality that accounts for this unity.

The holistic view of energy as a kind of "quick-change artist" with a whole trunkful of disguises was very much in the spirit of a cultural movement of the late eighteenth and early nineteenth centuries. The artistic, literary, and musical manifestations of the movement went by the name *Romanticism,* which we will use to denote the movement as a whole.

The romantic movement was a reaction to the extreme rationalism of the Enlightenment, to the sterile formalism of the music, art, and architecture of that period, and to the horrors of the new industrial society, with its "dark satanic mills." While romantics did not reject reason per se, they extolled the creative powers of emotion and intuition.

Romanticism had a scientific offshoot called *naturphilosophie.* The exponents of this school came from many branches of science, but put heavy stress on the organic wholeness of nature. Energy was their word for the vital principle behind all change, motion, growth, creativity, and passion. This movement was particularly strong in the German-speaking universities of central and northern Europe.

Without this sort of passionate conviction, the energy concept might never have made much headway, for in its early stages of development the law of energy conservation had to be taken pretty much on blind faith. Most forms of energy were little understood, so there were very few processes in which all the energy could be accounted for.

In retrospect, it is hard to say whether the holists won or whether they were simply co-opted. Though the energy concept grew to encompass a wide range of natural phenomena, truth, beauty, and wisdom remained beyond its scope. With the introduction of the atom, nearly all forms of energy came to be understood in reductionist terms. But before we move on to this topic, we must investigate one of energy's most mysterious guises, *potential energy.*

A MONEY-BACK GUARANTEE

All the transformations of energy we have considered so far involve some noticeable change in the world; gasoline is burned, brakes heat up, and so forth. But with certain forces, especially gravity, something mysterious happens. A falling stone picks up energy as it descends. Where does this energy come from? Nothing has apparently changed except the location of the stone, but energy has miraculously appeared.

This puzzle can be illustrated by the example of a stone lifted by a hoist, as shown in Figure 5-1. From the point of view of energy, this is the reverse of free fall, for energy is expended with no visible return other than a change in the height of the stone.

From the older Newtonian point of view nothing very interesting is happening. Through most of its rise the stone moves at constant speed, its weight exactly balanced by the force exerted by the rope. In terms of Newton's laws, the stone might as well be standing still. Yet the person turning the crank may be working up a considerable sweat. Work is going into the process, the product of the weight mg of the stone and the height h it is lifted. Yet this work does not go into increasing the stone's kinetic energy, nor does it disappear as heat. The counterbalancing force of gravity has taken it right out again. Is that work lost forever?

The answer, of course, is no. If the rope is cut, the stone will fall. Gravity will work on the stone, and the energy put into lifting it will be returned, mostly as kinetic energy. The stone reaches the ground moving at the same speed it would have acquired if the work done to lift it had not been opposed by gravity.

Thus, gravity seems to be an "honest" force; work done to overcome its effect may produce no immediate reward in the form of motion, but it can be recovered later. Not all forces have this nice property; the work done dragging a stone across rough ground is forever lost.

Energy "stored" in this form is called *potential energy.* The sense of the word is self-explanatory. By raising the stone we have created a situation that can potentially lead to motion. Allowing the stone to return to its starting point will convert that potential to an actual motion. The direction of the force is vertical, so only the vertical portion of the motion counts. The symbol V is customary for potential energy, and the formula for gravitational potential energy is $V = mgh$.

Except for a brief start up the forces are in balance

Afterward the stone rises at constant speed because the net force is 0

mg + "a little bit"

Thus, as the stone rises, there is no acceleration; yet work is being done by the man at the crank

↑mg
↓mg

But if the stone is released by cutting the rope at the top...

...it reaches the ground with the same speed it would have acquired had the work been done in the absence of gravity

FIGURE 5-1.

The process of converting potential energy into kinetic energy is a gradual one. When the stone has fallen only one-tenth of the way to the ground, gravity has done only one-tenth of the work it will finally do; one-tenth of the energy has become kinetic, the other nine-tenths remain potential. As the stone continues its fall, the potential energy shrinks and the kinetic energy grows. Ignoring all nonmechanical forms of energy, at all times the kinetic and potential energy must add up to the work done in lifting the stone. We can predict the speed of the stone at any point of its descent.

But Galileo could have done the same thing without introducing all these new concepts. If this were applicable only to a stone in free fall, it would hardly be worth the trouble. But consider the roller coaster depicted in Figure 5-2. The formula applies equally well to it. Once it is hoisted to the top of the first rise and released, it moves subject to only two forces: gravity and the support provided by the rails. But the support force is perpendicular to the motion, and therefore *does no work*. The speed acquired while dropping a given vertical distance is the same whether an object falls freely or goes down an incline! Galileo knew this too, but had to employ a far more difficult argument to prove it.

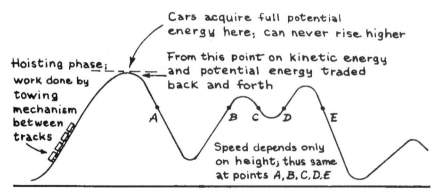

FIGURE 5-2.

Just as the roller coaster draws on its supply of potential energy as it falls, it puts energy back as it rises. Anywhere along the roller coaster's track, its speed depends solely on how high it is above the ground. The relationship can be expressed in a simple formula:

$$\frac{1}{2}mv^2 + mgh = \text{const.}$$

To find the value of the constant term, simply evaluate the expressions on the left-hand side of the equation using the height and speed of the coaster when it is released from its towing mechanism at the top of the first rise. To keep it constant, the higher h gets the smaller v must be. Of course for a real roller coaster there is a third force, that of friction, so an additional term must be added to the formula. This term grows continually throughout the ride, because energy removed by friction is never returned to the system. Then at each successive return to the same height the car is going a bit slower. The friction term depends on the speed and the distance traveled, and is a bit complicated to deal with here.

Gravitational potential energy has been a major energy source for the human race since before the dawn of civilization. From the primitive water wheel to the turbogenerators at the Grand Coulee Dam, we have exploited the potential energy of water as it descends to the sea. This is a self-renewing energy resource, because the ultimate source of water power is solar energy. Sunlight is absorbed in lakes and seas and converted into heat. This heat evaporates water, which is carried to the high clouds, to come down again as rain and keep the rivers flowing. Of course, damming this flow to tap its potential energy can have serious ecological consequences.

You may well be wondering at this stage where the energy *goes* while it is in the invisible "bank" of potential energy. If we stick to the strict action at a distance interpretation of gravity, the answer is "nowhere," and the mystery remains. But in Chapter 6 we will see that the field concept does give a sensible answer to this question.

ENERGY AND ATOMS

One of the greatest advances of nineteenth-century science was the discovery that if matter is regarded as being composed of atoms, all conversions of heat to other forms of energy could be understood in detail. Imagine, as in Figure 5-3, a collision between two balls of soft clay of equal mass, heading toward each other at equal speed; after the collision they will be fused together and standing still. Without even bothering to look into the internal structure of the clay balls, we find that total momentum is conserved, for momentum conservation holds at all levels. But the kinetic energy seems to have been lost.

But if we examine things on a deeper level, we find that the energy of motion was merely transferred to the atoms of which the clay balls are composed; this can be demonstrated by noting the increase in temperature of the clay balls, for temperature is a measure of the average energy of these atoms. This motion is complex, chaotic, and random, and there is no way to reverse the process; to get the atoms of a clay ball moving in one direction to restore the original energy of gross motion would be impossible.

When we go from the macroworld of clay balls to the microworld of atoms, the law of energy conservation becomes purely mechanical. Not only does heat yield to this analysis; chemical energy can be seen as potential energy of the forces that bind atoms together. This brought forth a hope that all natural phenomena might ultimately yield to a mechanical interpretation, if only their microscopic details could be understood. Then kinetic and potential would be the only remaining disguises for energy. Thus atomism, which will be the topic of the final six chapters of this book, allowed physicists to accept energy without deserting their deep-felt reductionist views.

To illustrate how the evolution of a science leads to new ways of understanding old problems, it should be noted that we now have yet a third way of understanding elastic collisions. From the Cartesian point of view, they arise from the mysterious property of certain objects to emerge from a collision with the same relative speed they had when they went in. Newton's laws allowed us to connect this behavior with a particular law of force. Now we see that, as with gravity, such a force must store potential energy during the instant of collision, and return it as the objects move apart. An elastic collision is thus one in which both momentum and *kinetic* energy are simultaneously conserved. If the collision is *inelastic*, energy is still conserved, but some of it is converted into other forms.

We can classify the three collision examples in Chapter 2 in terms of what kinds of energy transfer take place (see Figure 2.4). In the first example, where the balls stick together, part of the kinetic energy is converted into heat. In the elastic collision, all the kinetic energy is retained. The third example requires that a substantial amount of energy be converted from some other form into kinetic energy at the moment of collision, perhaps by attaching an explosive cartridge to one of the balls.

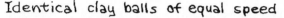

Identical clay balls of equal speed

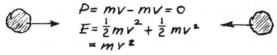

$$P = mv - mv = 0$$
$$E = \tfrac{1}{2}mv^2 + \tfrac{1}{2}mv^2$$
$$= mv^2$$

After collision, momentum
conserved but energy lost

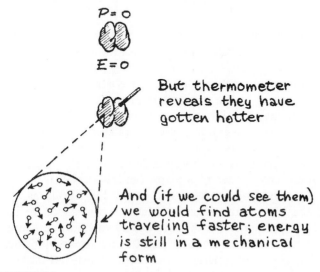

$P = 0$

$E = 0$

But thermometer
reveals they have
gotten hotter

And (if we could see them)
we would find atoms
traveling faster; energy
is still in a mechanical
form

FIGURE 5-3.

BINDING ENERGY

The potential energy concept is a very useful tool for dealing with situations in which objects are bound together by forces of attraction. The Earth and Moon are bound to one another in this fashion, and the Earth and its sister planets are bound to the Sun. In the microworld, atoms are bound into molecules by a form of electrical attraction.

From the energy viewpoint, such objects are bound because they do not have enough energy to escape one another. To move the Moon away from the Earth, one would have to put in energy in the form of work against their mutual gravity.

For example, if an object leaves the Earth's surface with a speed greater than about 11 km/s, it is free to "escape" the Earth's gravity. It still slows as it moves away, but gravity can never do enough work to bring it to a halt. No matter how far it goes, there will still be some energy left. This is the meaning of the term *escape velocity*. Escape would not be possible if gravity did not drop off as the square of the distance. Were the force to remain constant, for example, one could do sufficient work to stop the object and bring it back simply by going far enough. The formula *mgh* is valid only for short distances, where *h* is small compared with the Earth's radius.

In situations like this, it becomes convenient and logical to choose a scale for potential energy in such a way that an object barely free to escape has zero total energy. When this convention is adopted, all potential energies for forces of attraction are *negative*.* An object gains kinetic energy as it moves inward in response to an attractive force. Thus, its potential energy must decrease. In a chemical reaction that releases energy, the binding energy becomes more negative.

This gives us a convenient scheme for classifying the motion of objects under attractive forces. If the total energy is positive, i.e., the positive kinetic energy exceeds the magnitude of the negative potential energy, the object is free to escape. If the total energy is negative, it is bound. There is a distance at which the potential energy itself is equal to the negative total energy of the object. Since kinetic energy cannot be negative, it can never go beyond this point.

This is, of course, an arbitrary convention, a mere bookkeeping device. As we shall see in the next chapter, objects attract one another because they can draw upon an enormous positive store of energy called a *field*. The negative potential energy simply represents a "debit," a loan from this store.

As an example, consider the lunar missions flown by the Apollo astronauts. They approached the Moon from a great distance; with respect to its gravity, they had positive total energy. To become bound to the Moon in a lunar orbit, and then descend to the surface, they had to slow down by firing their rockets in reverse, "giving away" energy. The maneuver was not undertaken without some trepidation; once they had thus obtained negative total energy with regard to the Moon, they could not escape it without putting energy in. Had their rockets been unable to restart and boost them to a positive energy with regard to the Moon, they would have been trapped.

When we study relativity, we shall see that binding energy takes on a more concrete manifestation, a defect in mass.

Stars and planetary systems are formed from the collapse of immense, diffuse clouds of gas and dust. In this process, a great deal of gravitational potential energy is converted to heat. If enough mass is present, the temperature at the center becomes high enough to kindle thermonuclear reactions, and a star is born. The minimum required is about one-twentieth the mass of our Sun.

Almost inevitably, the cloud has some angular momentum. As it collapses, it must turn more and more rapidly, just as skaters or dancers create a faster spin by pulling in their arms. Eventually, the outer portions of the cloud reach orbital velocity and can fall no further. It is from this material that planets form. The fact that all the planets in our Solar System move in orbits in the same direction, and the Sun itself rotates in this direction, is a clear indication of this history.

---

*The formula for gravitational potential energy is then $V = -G\,\dfrac{mM}{r}$.

In many cases, one or more of the planets grows big enough to become a star in its own right. A substantial fraction of the nearer stars in the sky can be seen, through a powerful telescope, to consist of double stars bound in orbits. Even when one of the partners is too small to be seen by its own light, its effect on the motion of its senior partner can reveal its presence. It seems unlikely that many stars are truly solitary. They have either stellar partners or planets.

Still, this gives us no assurance that anywhere in the universe there is a planet as lush with life as our own. Even in our Solar System, it appears that Earth is the sole planet hospitable to life. Perhaps life is a matter of such delicate balance that it is extremely rare. Science cannot yet tell us whether or not we are alone in the universe.

Summary

Holism is an approach to science that seeks connections between the parts of nature, while *reductionism* seeks to understand the whole by studying the parts. The concept of *energy* originated as a holistic concept that linked motion to phenomena such as heat, electricity, and chemical reactions. It is connected to the Newtonian scheme through a nondirectional measure of motion, *kinetic energy,* and through a measure of energy transfer by the action of a force, *work.* The concept originated with holistic philosophers influenced by the romantic movement, and gained acceptance in mainstream science through the work of James Joule, who found the quantitative connection between motion and heat. The SI unit of energy is the *joule.* A useful auxiliary concept is *power,* energy transfer per unit time. Its unit, the *watt,* is 1 joule per second. Power can be seen to be the appropriate quantifier for "propulsion" in Aristotle's analysis of motion at constant speed in the face of resistance. Viewed on the atomic level, heat is a mechanical form of energy. Thus through atomism, energy acquires a reductionist significance. An elastic collision can be defined as one in which *kinetic* energy is conserved. Gravity and similar forces are capable of "storing" energy, giving rise to the concept of *potential* energy. This provides a basis for understanding systems of objects bound by mutual attraction, such as the Solar System. Gravitational potential energy plays a key role in the formation of stars.

CHAPTER 6

One Last Part for the Machine

With Earth's first Clay they did the Last Man Knead
And there of the Last Harvest sow'd the Seed:
And the first Morning of Creation wrote
What the last Dawn of Reckoning shall read.
—FitzGerald, The Rubáiyát of Omar Khayyám

The vision of the universe as one vast machine was not buried with Descartes. The triumph of Newtonian physics only whetted the appetite of the *mechanists* who pursued this dream. To their dismay, Newton had left them a machine from which one crucial part was missing. As long as gravity remained based on action at a distance, it was simply a law without an underlying mechanism. Though some Newtonians were quite content with this situation, a few tried to do something about it.

Gravity, however, does not readily lend itself to experimental science. On a laboratory scale, the force is too feeble to do much with. And the strong gravity of the Earth is simply *there*—it cannot be controlled or altered in any way.

For that reason, the breakthrough to a more complete understanding of forces that act at long range had to come through studies of two stronger and more controllable forces, electricity and magnetism. This research established the *field* concept, which mechanists eagerly embraced as their long-sought missing part. One unexpected consequence of this work was the solution to the age-old mystery of the nature of light. By the end of the nineteenth century, some physicists were convinced that the physical universe was now completely understood, and nature had no more secrets to disclose in that realm. Bright young students were advised to find something else to work on.

ELECTRICITY AND MAGNETISM

Electrical research began in earnest during the "enlightened" eighteenth century. Some of the most important discoveries were the work of scientific amateurs with little formal education. None of these contributed more than Benjamin Franklin.

Franklin was born in 1706, when Newton was the most revered living Englishman, and science was proclaimed the noblest calling to which an intelligent human being could aspire. Raised in Boston in a family of skilled craftsmen, he left school at the age of 10, the normal stopping point for someone in his circumstances. To escape from an onerous apprenticeship, Franklin fled to Philadelphia while still in his teens. He landed in the Quaker city practically destitute, but Philadelphia was then a rapidly growing boom town, the largest city in British America and the premier gateway for immigration.

Through hard work and political skill, Franklin quickly built an enormously successful printing business. With other young tradesmen he organized the "Leather Apron Club," which was devoted to civic works and the advancement of knowledge, as well as to rollicking good times. By the age of 43, he was well enough established to sell out to his partner and retire in modest comfort. Franklin had no intention to remain idle —he hoped to devote the rest of his life to science. Just two years later, in 1751, he published his *Experiments and Observations on Electricity.*

This work set the standard for electrical research for more than a generation. It was entirely nonquantitative, but its principles and terminology endure to this day. Electricity arises from two forms of "electrical charge" that he called *positive* and *negative*. Like charges repel one another, while opposite charges attract. Normal matter contains equal amounts of both signs of charge, so it is electrically neutral. Various physical and chemical processes, however, could destroy this balance. Here was a force stronger than gravity, and furthermore one that could be turned on or off at the experimenter's will.

Worldwide fame soon brought an end to Franklin's leisure to engage in full-time scientific work. The troubles with the mother country that ultimately led to the War of Independence had begun. His international reputation as a scientist made Franklin too valuable as a representative abroad for the American cause.

Franklin's example inspired the French engineer Charles Coulomb to put the study of electricity on a sound Newtonian basis through a law of electrical force. In 1789, Coulomb demonstrated that electricity, like gravity, obeyed the inverse-square law. To take the role played by mass in Newton's gravity, Coulomb used Franklin's electrical charge, so the force was proportional to the product of the charges divided by the square of the distance apart, a formula exactly like Newton's:

$$F = k\frac{qQ}{r^2}$$

where q and Q are the electric charges and k is a universal constant like that in Newton's formula.

As the nineteenth century dawned, Count Alessandro Volta of Como, Italy, developed the electrical battery, which could maintain a *current*, or steady flow, of electric charge. This opened a wide range of new experimental possibilities. In 1820, these bore fruit in the discovery of a connection between electricity and magnetism by a Danish professor, Hans Christian Ørsted, a *naturphilosoph* who was specifically looking for such a connection. He found that a compass needle will align itself perpendicularly to a wire carrying a strong electric current. A host of scientists rushed to exploit this discovery, but one young Englishman managed to get ahead of the pack and remain there for the rest of his life.

FARADAY AND THE FIELD

In London in 1812, a 21-year-old apprentice bookbinder presented himself to Humphry Davy, who had advertised for an assistant to help with his chemical researches. Michael Faraday's credentials consisted solely of a set of neatly illustrated bound notes taken from Davy's public lectures. That these notes were enough to land him the job was due in part to the ideology of the institution of which Davy was the director.

The Royal Institution had been founded for the express purpose of improving the lot of the British working class through science. It provided a laboratory for research to upgrade their standard of living, and also served as a beacon of light through evening public lectures. Faraday was one of the very few actual manual workers to attend; work weeks of seventy hours left little time for "self-improvement." But the London middle class found charities of this sort less burdensome to the pocketbook than decent schools and a living wage. As an embodiment of the *raison d'être* of the institution, Faraday could hardly be dismissed out of hand.

It soon became obvious that Faraday was a talented researcher in his own right. He gradually won his independence from Davy and, at the age of 34, succeeded him as director. Shortly thereafter, he abandoned chemistry and turned his hand to electrical research, following up on Ørsted's discovery.

Though he had no more formal schooling than Franklin, Faraday was by no means unsophisticated. He had been active for some time in a reading circle of young men in situations similar to his own. Through independent study, he had become extremely well versed in all branches of natural philosophy. He was much influenced by the writings of the Jesuit Father Rudjer Boscovich, a native of Ragusa (now Dubrovnik, Croatia) on the Adriatic coast.

Boscovich, a contemporary of Franklin, had argued that in the Newtonian scheme there was no longer any need for separate concepts of *force* and *matter.* The ultimate atoms of matter might well be nothing more than points that served as centers of force. This idea is now central to our current picture of the nature of matter, and we will return to it in the final chapter of this book. But in its time, it was no more than an untestable speculation. Faraday, however, found in it something of immediate utility. If force was to be the ultimate reality, he was sure that it must be based on something more substantial than action at a distance.

Faraday believed that long-range forces worked by filling the space around objects with something he called a *field*. Each object contributes to the field, and responds to the combined field of all of them. As an aid to visualization, he developed a pictorial scheme called *lines of force*, illustrated in Figure 6-1. These lines represent the field in two ways: the direction of the force at any point in space is along the lines, and the strength of the force is greatest where the lines are closely spaced.

To express this in a formula, we define something called the "field strength" E, and Coulomb's law becomes *two* formulas:

Faraday and his wife.
(*The Royal Institution.*)

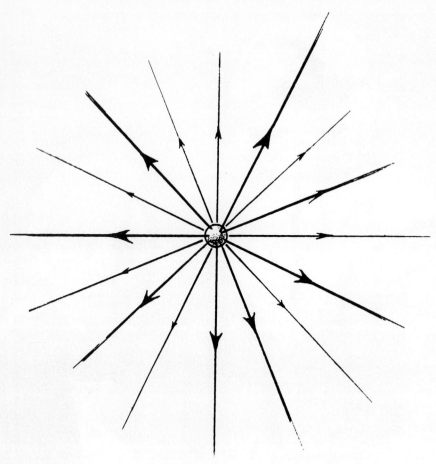

FIGURE 6-1.

$$E = k\frac{Q}{r^2} \qquad F = qE$$

The power of this concept is not very evident when dealing with a single charge. The lines of force produced by a system of two opposite charges is depicted in Figure 6-2. In this situation, one of the rules for drawing these lines of force is that each one begins on one charge and ends on the other. Lines of force become even more useful when we move from electricity to magnetism, a complicated force that is not a simple matter of attraction or repulsion. Magnetic fields are generated only by moving charges; if the charges are standing still, there is no magnetic field. The simplest case is the magnetic field of a steady electric current, shown in Figure 6-3. The lines of force do not radiate out from a current-carrying wire, but form rings around it. Ørsted's compass needle aligned itself with these lines of force.

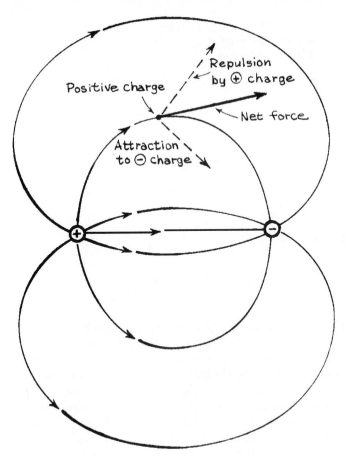

FIGURE 6-2.

The magnetic field also only affects moving charges. The force on a charge q moving at velocity v at an angle θ with respect to a magnetic field of strength B is:

$$F = qvB \sin \theta$$

Furthermore, this force is perpendicular to both the direction in which the particle is moving and the field! It is awkward to express such complicated geometric relations as a simple matter of action at a distance, so in magnetism the field concept truly comes into its own.

In a brilliant series of experiments, Faraday discovered that the relation between electricity and magnetism is completely reciprocal. A *moving or changing electric field generates a magnetic field, and a moving or changing magnetic field generates an electric field.* With this intimate a connection, it made sense to think of the two fields as different forms of a single *electromagnetic* field.

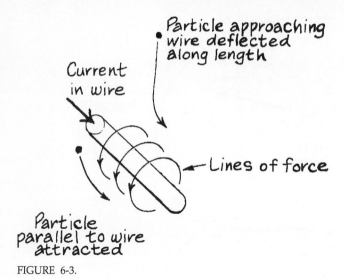

FIGURE 6-3.

An electrical conductor is a material in which electrons, which carry electric charge, are free to move. When a conductor moves through a magnetic field, the force on the electrons causes them to move, generating an electric current. This is the principle underlying the *dynamos* we now use to generate electricity. Faraday built a crude prototype dynamo, which could not yet compete with batteries as a source of electric current. On a visit to the Royal Institution, Chancellor of the Exchequer John Peel asked him, "Of what use is it?" Faraday's reply is now classic: "One day, sir, you will tax it!" Within a generation, the descendants of Faraday's dynamo were beginning to light up the world.

It must be emphasized that fields and action at a distance have exactly the same observable consequences as long as the field is constant in time. Thus while Faraday's discoveries were widely admired, his fields were not taken very seriously until his work was refined and extended by a very sophisticated mathematical physicist, James Clerk Maxwell.

THE FIELD IS REAL

Maxwell, born to the minor gentry of Scotland, had begun his scientific career as a teenage mathematical prodigy. But he also had a bit of the tinkerer in him, with a love for mechanical invention and a well-developed physical intuition. So while he pursued physics on the highest levels of mathematical abstraction, he had a deep respect for the intuitive depth of Faraday.

Examining Faraday's discoveries in order to embellish them with a proper mathematical formalism, Maxwell discovered a startling implication: the transfer of momentum and energy via electromagnetic fields is *not instanta-*

neous. There is a time delay, equal to the time it would take light to pass from one object to the other! This brief delay was far too short to observe in the laboratory; only mathematics could reveal its presence.

When objects interact with one another via electromagnetic fields, each is continually exchanging energy and momentum with the field. Some or all of this momentum and energy may end up on the other object, but only some time later. If energy and momentum conservation are to survive, energy and momentum must be credited to the *field itself.*

This put a more solid foundation under the concept of potential energy. Maxwell showed that wherever a field exists, energy is distributed throughout that region of space. It is this bank of energy on which an object draws when it converts potential energy into kinetic energy. Negative binding energy arises from the fact that when two objects that attract one another are close together, their combined field contains less energy than their separate fields did when they were far apart.

Maxwell also found the speed of electromagnetic force transmission terribly suggestive. Perhaps *light itself* could be some arrangement of electromagnetic fields. He quickly discovered a pattern that would do the trick. An electric and magnetic field are at right angles to one another and in motion. The moving electric field generates a magnetic field, and the moving magnetic field generates an electric one. If they move at the speed of light, the two fields exactly sustain one another, without any need for an outside source. Light is such an arrangement, in a repeated wave pattern as shown in Figure 6-4. We will learn more about these waves in Chapter 7.

There was a clear implication in this work that other kinds of electromagnetic waves might exist. Within a few years, these were discovered in Germany by Heinrich Hertz. Modern radio and television are based on these waves. So a revolution in communications, and a worldwide industry, can be traced to the effort to find mathematical unity in electricity, magnetism, and light.

Once light or any other electromagnetic wave is sent on its way, it can travel across the Earth—or even across the universe—until something gets in its way and absorbs it. Its existence no longer depends on the electrical charges and currents that produced it. Faraday's fields had now acquired momentum, energy, and an independent existence. What more need one demand to call them *real!*

Maxwell's ideas had implications that went beyond electromagnetism and light. It seemed likely that any fundamental force, such as gravity, might act through a field. If this field did not propagate instantaneously, but at a finite velocity like that of light, then it too must produce some form of radiation to carry the momentum and energy when the field is somehow changed. In Chapter 12, we will see how this insight encouraged Albert Einstein to invent the field theory for gravity that brought him worldwide fame.

Maxwell's electromagnetic theory was to be the high-water mark of Newtonian physics, its last great triumph. His principal work, the *Treatise on Electricity and Magnetism* published in 1873, rivals the *Principia* in its significance.

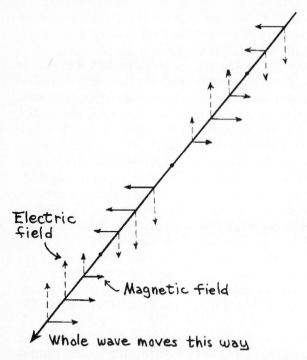

Electric field

Magnetic field

Whole wave moves this way

FIGURE 6-4.

To the mechanists, only one seemingly small step remained, to find a mechanical interpretation of electromagnetism. Maxwell tried to do this by imagining a substance called the *aether*, which would fill all space. This was not a new idea—light was known to have a wave nature, and there had been a great deal of speculation about what medium light traveled in. But Maxwell hoped his mathematical laws would nail down the properties of the aether precisely.

Electric and magnetic fields would be strains and flows of the aether, while light would move through it as sound waves move in air. But the aether had too many contradictory properties. It had to be as rigid as a solid, yet flow like a liquid, while never impeding in any way the motion of the Earth and the planets. Even worse, Maxwell found that he could not tie down a single unique set of properties. And if he wanted to account for gravity as well, he would need a different sort of aether entirely. Though he remained convinced of the reality of his aether, he reluctantly concluded that it was not yet good solid science, and left it out of his *Treatise.*

It was a wise decision, for the attempt to install this one last part in Descartes' universal machine was to end the long reign of Newtonian physics.

The careers of Faraday and Maxwell spanned the period when scientific research became a profession. When Faraday was hired at the Royal Institution, he became one of only a handful of people in the entire world who were paid a salary primarily to do scientific research. But starting in Germany in the middle of the nineteenth century, the practice of founding research institutes affiliated with universities spread rapidly. Maxwell was to spend his later years as the first director of one of the most eminent of these, the Cavendish Laboratory at Cambridge University. Most of the support for this laboratory came from the nascent electrical industry, and the first order of business was to improve the accuracy of electrical measurements.

At these institutes, young scientists learned research by the oldest and most effective form of teaching yet devised, apprenticeship under a master. In most nations, the students were working toward the Ph.D. degree, but at Oxford and Cambridge the English prejudice that "gentlemen don't dirty their hands" delayed this form of certification for several decades.

THE NIGHTMARE OF DETERMINISM

The spectacular rise of Newtonian science led many thoughtful people to give considerable credit to its claims of universal validity. This view of reality ultimately led to the conclusion that everything that happens in the universe is a consequence of the motions and interactions of atoms.

In Newton's physics, motion is governed by perfectly deterministic laws. Early in the nineteenth century, the mathematical physicist Pierre Simon de Laplace speculated that if one could only observe at some instant all the atoms in the universe and record their motions, both the future and the past would hold no secrets. Put another way, all of history was determined, down to the last detail, when the universe was set in motion. The rise and fall of empires, the passion of every forgotten love affair, represent no more than the inevitable workings of the laws of physics; the universe marches to its unalterable destiny like one gigantic clockwork. The poetic expression of this vision in the *Rubáiyát* serves as the epigraph of this chapter.

What room did this leave for free will, for salvation and damnation, for love and hate, when the most trifling decision any human being can make was determined more than 10 billion years ago? It gave the ethical thinkers of the nineteenth century something to ponder. Admittedly, it is inconceivable that one could actually achieve the omniscience required by Laplace. But the fact that it was possible in principle was viewed as a genuine nightmare.

This sort of social determinism was exemplified in the tactics of the celebrated attorney Clarence Darrow. Defending a client who was patently guilty of the offense as charged, Darrow would point to him as the prisoner of his own heredity, placed in an environment not of his own choosing. Under such circumstances, following from a chain of causes leading back to time immemorial, what was the meaning of "responsibility for one's actions"?

Even those untroubled by this admittedly abstract nightmare felt the impact of physics. For the first time, all the details of a tremendous range of natural phenomena were understood in terms of a few simple principles. Newtonian physics became a model that all of human knowledge should aspire to. As the social sciences began to emerge, they tended to distance themselves from the humanistic studies from which they had sprung. Social thought turned reductionist, seeking general laws to explain history and human behavior. In the wake of this movement, figures such as Karl Marx and Sigmund Freud have had a profound impact on history.

DOES CHAOS RULE THE UNIVERSE?

It is important to remind ourselves that faith in determinism rests on one achievement that was without precedent in science, and that has not been duplicated since. Newton's derivation of Kepler's laws seemingly described the Solar System as it has existed for all time, and will exist into an unending future. But Newton himself was fully aware that this could not be the whole story. Kepler's laws apply perfectly only in a Solar System ruled by the Sun's gravity alone. They take no account of the forces that the planets, through their gravity, exert on *one another*.

There is a fundamental reason for this omission. There is no simple exact mathematical solution for the motion of more than two objects interacting with one another. This was true in Newton's day and remains so today. Kepler's laws work because the Sun is much heavier than any of the planets. Jupiter, the largest, is a thousand times lighter than the Sun. Still, the Earth receives, over thousands of years, transfers of momentum from Jupiter that equal in magnitude the effects of the Sun's gravity over one year. Thus it would not be surprising to see major changes in the Earth's orbit on a time scale of thousands of years.

Newton considered this problem and found it reassuring rather than alarming. He secretly had little use for the remote God of his deist friends, preferring an Old Testament deity, involved in the day-to-day management of His creations. The solar system would be kept stable by the direct intervention of a benevolent Lord.

Laplace later showed that the mutual attractions of the planets tend to average out, and the instability that Newton feared amounts to a number of slow, cyclical variations of the planetary orbits. But these were only approximate calculations. Later in the nineteenth century, the French mathematician and philosopher Henri Poincaré addressed the general question of the mutual interaction of just three bodies, and found that some arrangements were highly unstable. Indeed, if one of the bodies is much lighter than the other two it is very likely to simply be ejected from the system. He was prescient enough to realize that, given the complexity of the real world, unpredictable situations must be far more common than predictable ones, and that in practical terms most of what happens in the world is beyond our power to predict.

Confessing that thinking about these problems actually made him ill, Poincaré abandoned them. Given the paper-and-pencil tools available to the theorist in that era, he had little choice. Today, cheap and powerful computers enable scientists to study systems far more unstable or complex than those that so upset Poincaré. The manuscript for this edition of *Physics for Poets* was written on a modest desktop workstation with more than the combined computing power of all the computers in the world at the time the author recieved his Ph.D. Studies with such machines have revealed just how unpredictable our world can be.

Even a system as simple as Poincaré's three bodies has a property that today goes by the name of "chaos." In such a system, a tiny, unmeasurable change in the starting conditions grows—sometimes rapidly—with time. This leads to a radical difference in the final consequences.

Today we realize that the motions of the two outer planets of the Solar System may be chaotic. This is hard to recognize because they are so far from the Sun that they move very slowly. Their orbits may be unstable, but it will take hundreds of millions of years for them to go seriously awry.

In the 1960s, weather forecasters turned to the computer as the answer to their hopes of better long-range predictions. The atmosphere obeyed physical laws that were well understood, but it was so large and complicated that only a super computing machine could hope to track its future development. In the years since then, the power of computers has increased by more than a hundred thousand times, and satellites now feed them ever-more detailed weather information. Yet the predictability of local weather remains locked at a five- to ten-day limit, because the atmosphere is chaotic. It has been suggested that just one flap of a butterfly's wings in a sensitive location might well determine whether weeks later and thousands of miles away a tornado will crash through a crowded residential area, or spend itself harmlessly on a barren plain.

Today, we have come to realize that there are limits to our ability to foresee the future. Some things, such as planetary motions, can be predicted for millenia. Others are good for a matter of hours, some only for fractions of a microsecond. The nightmare of determinism is exactly what the word implies—a bad dream with little connection to reality. Any small defect in our knowledge of the present can lead to drastic changes in our vision of the future. In Chapter 17, we shall see that quantum theory has shown that we can never know the present perfectly. The future, just as common sense and the old popular song "Que Sera" tell us, is not ours to see.

TOWARD A SCIENCE OF COMPLEXITY

In the latter years of the twentieth century, the computer has become the holist's best friend. Though chaos rules out the possibility of predicting all the *details* of the future of anything as complicated as the world we live in, it does allow us to discern certain stable patterns that are likely to occur. Computer simulations of very complicated situations both inside physics and in other realms, such as ecology or economics, have shown that complicated systems have a way of generating their own order.

For example, ecologists have done repeated simulations in which a hundred species are placed on an imaginary "island" and allowed to interact. As time progresses, species go extinct. Eventually, a residue of twenty or so species achieves a stable equilibrium. But if they start the process over again, there is no assurance that they will find the same pattern of surviving species. There are many stable patterns, and which one arises can be largely a matter of chance, but once it establishes itself it can last almost indefinitely—unless there is some radical change in conditions.

One such radical event was the cometary impact that ended the reign of the dinosaurs some 70 million years ago, paving the way for the dominance of mammals as the large terrestrial animals. As a result of this catastrophe, the worldwide ecosystem went from one state of comparative equilibrium to another. Were it not for that event—completely unpredictable given the chaotic motion of such minor bodies in our Solar System—neither the author nor the readers of this book would be here today. It is sobering to contemplate that this was not the first event of its kind in Earth's long history, and is unlikely to be the last.

Nineteenth-century evolutionists saw the emergence of the human race as the inevitable consequence of the ascending evolution of life toward higher stages of perfection. Today, this comfortable notion is hardly tenable. It seems more likely that we are the rather arbitrary—and temporary—consequence of a long train of historical accidents.

Though the future may never be predictable in detail, there is some value in studying the "islands of stability" that emerge in complex systems. Such studies now go by the name *complexity theory*. This is cross-disciplinary science, embracing physics, embryology, ecology, evolution, and even economics. Its mecca is the Santa Fé Institute in New Mexico.

Complexity studies tend to focus on "emergent" qualities, features of a system that are consequences of the interactions of its parts, rather than inherent in their nature. One familiar example is the doctrine of the "invisible hand," promulgated in the eighteenth century by the Scottish economist Adam Smith. In a free economy, individuals pursuing their own selfish ends create a market that serves the common good. Interestingly, complexity researchers have found that the invisible hand is not what they call a "robust" emergent quality. It easily falls victim to monopolies, price-fixing agreements, and other restraints on trade. This coincides with real-world experience, in which we find we need the Securities and Exchange Commission, antitrust laws, bankruptcy courts, and other artificial stabilizers to keep free markets healthy.

But by far the most outstanding example of emergence is the phenomenon that is at once the most familiar and most mysterious thing we know—our own consciousness.

No modern scientist seriously doubts that everything that happens in the human brain is a consequence of the physics of the atoms and molecules of which it is formed. Nonetheless, few have enough faith in reductionism to suggest that physics alone will ever explain how this amazing bundle of subatomic particles becomes *self-aware*. The trillions of interconnections within one human brain dwarf the millions found in the most powerful computers ever made. There are far too many connections in the brain to program in our genes. The brain organizes *itself* as we grow and learn. But from the earliest moments of life there is the amazing awareness of self, expressed in Descartes' most famous assertion: "I think, therefore I am."

Unflinching reductionism still rules the roost in some of the proudest bastions of science, ranging from neoclassical economics to particle physics, and most notably in molecular biology, where the fruits of reductionism have been a revolution in science and technology. Nonetheless, holistic ideas are on the march, and it would not be surprising to find holism a significant theme in the science of the twenty-first century.

Be that as it may, one thing is abundantly clear: modern science has finally laid to rest both the mechanistic dream of Descartes and the deterministic nightmare of Laplace. We can never again pretend to understand our universe as a mere machine, regular and unerringly predictable over the vast span of time.

Summary

The one missing piece in a completely mechanical picture of the universe was a mechanical interpretation of forces like gravity, which act at a distance. An attempt at this was provided by the concept of a *field*, which distributes energy in space. It was developed not in connection with gravity, but through studies of electricity and magnetism. Benjamin Franklin made significant contributions to the understanding of electricity, which proved to be like gravity, an inverse-square force. The connection between electricity and magnetism was discovered by Hans Christian Ørsted, and studied in detail by Michael Faraday, who found that a moving or changing magnetic field produces an electric field, and vice versa. A mathematical formulation of these ideas by James Clerk Maxwell revealed that a field is not transmitted instantly, but at the speed of light. Thus light came to be understood as an electromagnetic wave, and the new technology of radio transmission, using longer electromagnetic waves, followed as a consequence. The attempt to find a mechanical model for a field led to the concept of an all-pervading fluid called the *aether*. Though physics at this point seemed nearly complete and wholly deterministic, modern studies of *chaos* have revealed this to be an illusion. Through computer studies of complex systems, however, patterns of order within this chaos can be studied.

CHAPTER 7

Waves

There is something fascinating about science. One gets such wholesale returns of conjecture out of such trifling investments of fact.
— MARK TWAIN, Life on the Mississippi

While the planets were the key to the triumphs of seventeenth century physics and astronomy, in the twentieth century *waves* were to play this role. So in this chapter we will break off our historical narrative to introduce a few simple wave concepts, the terminology used to describe them, and two significant effects that play a dominant role in contemporary physics.

A wave is not a material object, but is instead a *pattern that moves*. As a water wave sweeps across a lake, the water does not move with it, but simply bobs up and down as the wave passes. The word *wave* has recently been used to describe a crowd activity now fashionable at sports events, and this use of the word is entirely appropriate. Spectators simply rise and lift their arms when the wave reaches them, while the wave itself sweeps around the stadium much faster than any human being could run.

Wave patterns can arise in a wide variety of circumstances. They can be deformations of a music string or bumps on the surface of a body of water. Sound waves are small variations in the pressure of air, and light and radio waves are patterns in an electromagnetic field.

Nearly everything important about waves can be understood in terms of two simple principles. These are not deep natural laws like Newton's, but Aristotelian-style generalizations that are not exact in all circumstances. These principles are:

1. Waves move at a constant velocity that is determined by the medium that supports them, rather than the waves themselves.
2. Waves obey a superposition principle: If two or more waves arrive simultaneously at the same place, the resulting effect is simply the sum of the effects of each of the waves.

Exceptions to the first principle include water waves, in which the dimensions of the wave influence its velocity. The second principle is violated when waves are so strong that they alter the medium in which they travel. Violent shock waves from explosions have this property. But there are many examples of waves that follow them almost perfectly, such as sound waves in air and light waves in a vacuum.

As simple as these principles are, they give rise to some effects that are far from trivial. For our purposes, the two most important are *standing waves* and *interference patterns.* The main goal of this chapter is to acquaint you with these effects.

MOVING BUMPS

The simplest examples of waves are single wave pulses traveling on a one-dimensional medium, such as a string. Left to itself, a taut string will remain straight. But if we pluck it near one end, we create a deformation, the "bump" illustrated in Figure 7-1. The tension in the string immediately acts to eliminate the bump, but Newton's third law prevents it from disappearing. If the portion of the string to the right of the bump pulls it down, it must in turn be pulled up. When the part of the string we plucked returns to its normal position the bump has not disappeared, but has simply moved to the right. As this process continually repeats itself, the bump keeps moving along the string.

Note that it is the bump that travels, not the string, which simply returns to its original position. The speed at which the wave moves depends on the tension and the mass of the string; the higher the tension, or the lighter the string, the faster the wave will move.

To show the versatility of waves, let us look at another example totally devoid of any connection with physics. Imagine a marching band formed in a long single line. Each member of the band is given the instruction, "Watch the players on either side of you—if one of them moves, you do the same thing on

FIGURE 7-1.

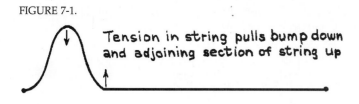

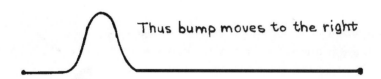

the next beat of the music." We then go to the end of the line and ask the musician there to take two steps out, then two steps back.

The resulting effect is illustrated in Figure 7-2. As viewed from above, a bump travels down the line of musicians from left to right, yet not one of them has moved either left or right. This is a true wave phenomenon in the full meaning of the word, except that the "medium" is not continuous, as it is in most cases. Nonetheless, our two principles can be applied in exactly the same way they are to more physical waves. To get superposition, we tell our players that if the neighbors on both sides move on the same beat, simply execute both movements on the next beat. If they move in opposite directions, simply stay put.

Again, the speed of the wave is set by the medium. If the musicians stand five feet apart and the music has two beats per second, the wave will move ten feet per second.

Let us return to the example of the string to illustrate wave superposition. In this one-dimensional case, two waves can meet only if they are moving in opposite directions, as in Figure 7-3. To distinguish the two waves, we make one much bigger than the other. The small wave is simply a moving bump on the big wave as they pass, and neither is altered by the encounter.

FIGURE 7-2.

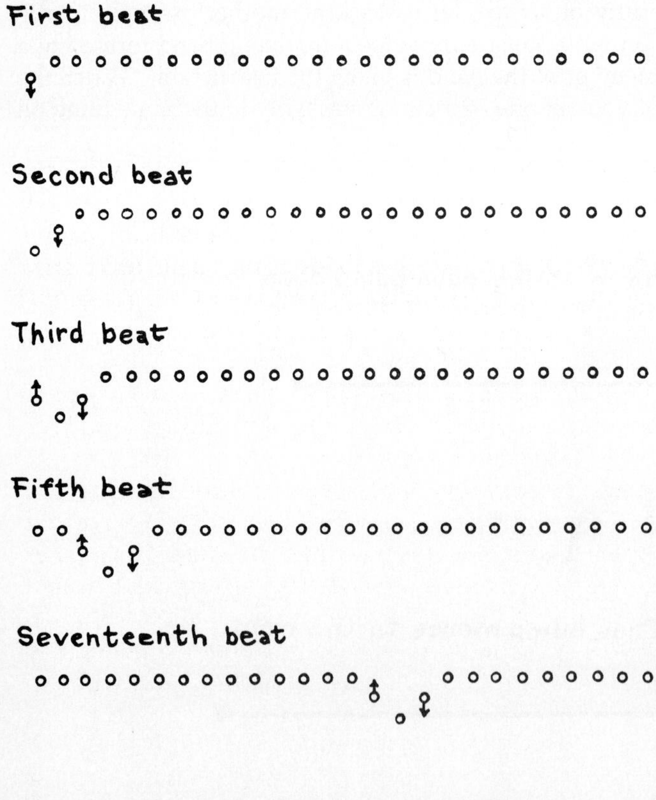

First beat

Second beat

Third beat

Fifth beat

Seventeenth beat

Before meeting

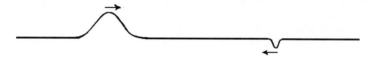

When meeting

Afterward

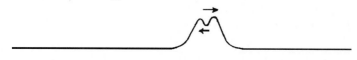

FIGURE 7-3.

The example gets more interesting if the waves are the same size and shape. In Figure 7-4 we see two versions of this, one where the bumps point in the same direction and one in which they are opposite. In the first case, at the instant they pass we simply have a wave that is twice as big as the individual waves. In the second case, there is a brief instant when the string is absolutely flat. But segments of the string are moving at this instant, and they will over-shoot, recreating the two individual waves.

This phenomenon goes by the name *interference*. When the waves rein-force, the interference is called *constructive*, and when they cancel it is *destruc-tive* interference.

Some of the more interesting applications of the principle of superposition are the "backward" ones. These are the cases in which we analyze a wave and predict its future development by decomposing it into the sum of two other waves, much as Galileo analyzed projectile motion by decomposing it into horizontal and vertical motions.

As an example, let us ask what happens if we form a bump in the middle of a string. It is equally free to move in both directions, with no inherent ten-dency to go one way or the other. What will it do?

(a) Superposition of identical pulses

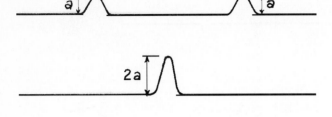

(b) Superposition of opposite but equal pulses

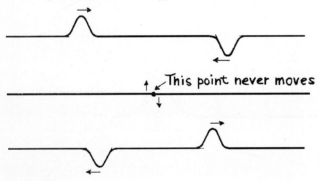

This point never moves

FIGURE 7-4.

Note that this situation is exactly the same as we had in the first example in Figure 7-4, at the moment of constructive interference. There is no difference in the shape or motion of the string in the two situations. Even though in one case the bump formed as the fusion of two waves and in the other we formed it ourselves, there can be no difference in the subsequent behavior of the wave. We thus predict that the bump will split into two waves, each of half its height, moving off in opposite directions, and observation verifies that this is the case.

In two or more dimensions a wave will spread in all directions, the pattern of ripples that form when we drop a stone in a pond, as illustrated in Figure 7-5.

PERIODIC WAVES

Single wave pulses are easy to follow, but are not very interesting. More challenging possibilities arise when there are trains of repeated, identical waves.

(a) Sectional view

(b) View from above

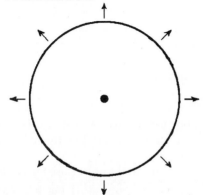

FIGURE 7-5.

Such waves are called *periodic*. They follow the same principles as individual wave pulses, so we can deal with them by simply adding a bit of descriptive terminology.

Two of the common terms are illustrated in Figure 7-6. The *wavelength*, for which we use the lowercase Greek letter lambda (λ), is the interval at which the pattern repeats. It is measured in units of length. The *amplitude* measures the size of the displacement produced by the wave. Here the units depend on the type of wave. For a water wave it is simply the height, but for a radio wave the amplitude would be the maximum electric field strength.

One more word is required to complete the description: since the wave is moving, any point on the medium goes through a motion that repeats itself as each wave passes. The number of times per second this happens is called the *frequency*, usually denoted by the lowercase Greek letter nu (ν). Frequency is measured in cycles per second, renamed *Hertz* (Hz) in honor of the discoverer of radio waves.

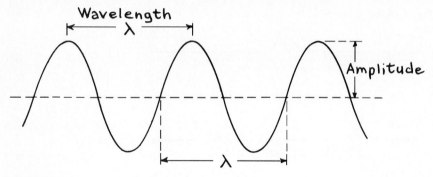

FIGURE 7-6.

Sometimes, in place of wavelength, it is more convenient to use the wave number k, the number of waves per unit length. This is simply the inverse of the wavelength; that is, $k = 1/\lambda$.

The wavelength and frequency are closely related, because the wave travels at a constant speed. For example, if a wave has a frequency of 5 Hz, five waves pass each second, and if each is 4 m long, the wave must be traveling 20 m/s. This relationship can be summarized in the formula

$$c = \lambda v$$

where c is the conventional symbol for the wave velocity. This is not a scientific law in the usual sense, but merely a relation that follows from the definitions of wavelength and frequency.

The velocity of sound depends on the air temperature, but on a typical day it moves at about 340 m/s (765 mph). Light and other electromagnetic waves are much faster, moving 300,000,000 m/s, or 300,000 km/s. Audible sound waves have frequencies ranging all the way from about 20 Hz to 18,000 Hz, and thus have wavelengths that range from about 2 cm to nearly 20 m. Light waves are extremely short and span only a narrow range of wavelengths, from 0.4 to 0.7 micrometer (μm, millionths of a meter). Their frequencies are extremely high, around 5×10^{15} Hz (5 quadrillion Hz).

The smooth wave shown in Figure 7-6 is called a *sine wave* because its mathematical description uses the trigonometric sine function. A sound wave that has this shape is heard as a pure musical tone whose pitch is determined by the frequency. A sine wave of light gives a pure spectrum color. Waves, however, can have almost any imaginable shape. As long as the shape is faithfully repeated for many wavelengths, waves can be built up by combining sine waves of different wavelengths.

STANDING WAVES

When a wave is confined between fixed boundaries, such as the ends of a music string, there are severe restrictions on the kinds of wave motion allowed. The wave reflects off both ends of the string, so there is no net tendency to move in one direction or another. Instead, we get patterns that look like waves standing still, one of which is shown in Figure 7-7. All musical instruments generate their sounds in some such fashion.

A standing wave can exist on a string only if it fits in such a way that the ends do not have to move. Since a wave pattern crosses the centerline once each half wavelength, the only waves that survive are those for which 1, 2, 3, . . . half wavelengths fit evenly into the length of the string. The case of three half wavelengths is shown in Figure 7-7. The points that do not move are called the *nodes* of the wave. The Principle of Superposition allows several patterns to coexist on the same string, so the actual pattern of motion can get a great deal more complicated than these simple ones.

FIGURE 7-7.

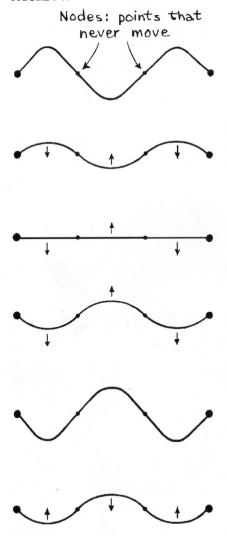

Since shorter wavelengths mean higher frequencies, the pattern with one half wavelength, called the *fundamental* or *first harmonic,* has the lowest frequency. The shorter waves are called *overtones* or *higher harmonics.* Real musical instruments are deliberately designed to produce a rich mixture of the fundamental and overtones, because a pure sine wave is a rather uninteresting sound.

A radio transmitter sets up a standing electrical wave in its antenna. A laser generates a standing light wave, confined between mirrors at either end, with a mechanism for pumping energy into the wave. When we get to quantum theory, we will find that standing waves play a crucial role in the structure of atoms.

INTERFERENCE PATTERNS

When periodic waves arrive at the same place from two synchronized sources, or from the same source by traversing two different paths, they produce an *interference pattern.* These patterns can be both strikingly beautiful and terribly useful. They provided a way to establish the wave nature of light, and also were the basis for a historic experiment that was crucial in the development of relativity. That experiment is the topic of the next chapter.

An interference pattern with sound waves is illustrated in Figure 7-8. Two speakers are sounding the same signal, an uninteresting pure tone. At the speakers, both sound waves are perfectly in step. But most places in the room in front of them are closer to one speaker than the other, so the waves no

FIGURE 7-8. Inference of sound waves from two speakers.

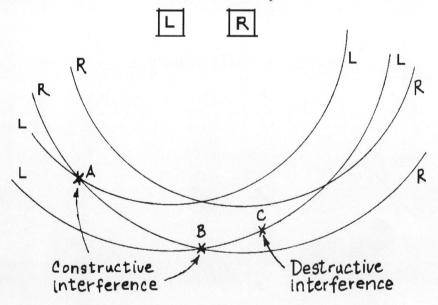

longer arrive perfectly synchronized, since they have traveled different distances to reach their common destination.

Point A in the figure is exactly one wavelength farther from the right speaker than from the left one. The waves from the two speakers arrive exactly one wavelength out of step. But since periodic waves are all identical, that is the same as arriving *in step*. The interference here is *constructive*. The waves reinforce one another, and a strong tone is heard. The same condition applies at point B, which is equidistant from the two speakers.

Point C, however, is one-half wavelength closer to its nearest speaker. Here the waves arrive exactly out of step. Maximum air pressure for one wave coincides with minimum air pressure for the other. In this case we get *destructive* interference. The waves cancel one another, and little or no sound is heard.

The key to understanding an interference pattern is straightforward: *take the difference between the distances from the two sources, and divide by the wavelength*. The resulting quotient q is given by the formula

$$q = \frac{L_1 - L_2}{\lambda}$$

The value of q tells you what kind of interference will take place. If it is an *integer* (i.e., -1, 0, 1, 2, 3, . . .), the interference will be *constructive*. If it lies halfway between two integers (i.e., $-\frac{1}{2}$, $\frac{1}{2}$, $1\frac{1}{2}$, $2\frac{1}{2}$, . . .) the interference is *destructive*. Intermediate values will give intermediate results: Quotients close to an integer give not-quite-perfect reinforcement, while those close to a half integer give not-quite-perfect cancellation.

For example, in Figure 7-8, if we take L_1 to be the distance from the left speaker and L_2 the distance to the right speaker, at the points labeled A, B, and C, then $q = -1$, 0, and $+\frac{1}{2}$, respectively.

It is hard to sychronize two light sources, so interference patterns with light are generally produced by splitting a light beam into two parts and recombining them on a screen. This effect was used by Thomas Young in 1789 to settle the long-standing controversy over the nature of light, which dated from the time of Newton and Huygens.

Newton pioneered in the study of light. It was he who proved that white light is a mixture of all the colors of the rainbow. His book *Opticks*, though not as Earth-shaking as the *Principia*, was a scientific milestone. He personally found it congenial to think of light as a hail of tiny particles. Huygens favored a wave theory, and his *Treatise on Light* was on a par with *Opticks*. Young was, at various times in his life, a professor of physics, a practicing physician, and an amateur philologist who played a significant part in cracking the code of the Rosetta Stone, the key to the Egyptian hieroglyphics.

Young's apparatus was terribly simple, but adequate to the task. He blackened a small square of glass with soot from a candle, until it was opaque. With a razor guided by a straight edge, he scribed two thin parallel grooves in the

FIGURE 7-9. A two-slit interference pattern.
(Courtesy of Brian J. Thompson, The Institute of Optics, University of Rochester.)

soot, as close together as he could, because he knew light waves must be very short. In a darkened room, he allowed a beam of sunlight to strike the glass square.

Light passed through the two slits, and by the time it reached a sheet of paper held a few feet away the beams had spread out enough to overlap. What Young saw was the striking pattern illustrated in Figure 7-9. The bright bands represent constructive interference, while the dark ones are destructive. The bands are closely spaced, because light waves are so short that one need not move very far along the screen to change the difference in path lengths by a half wavelength.

Young's experiment was not all that new. Interference effects had been observed before—one was even discovered by Newton himself! But such was the prestige of Newton's name that many physicists clung to the particle theory until well into the nineteenth century. We will see later that this was not entirely foolish.

THE ELECTROMAGNETIC SPECTRUM

Today, we have developed technologies for handling electromagnetic waves of nearly all imaginable wavelengths and frequencies. We will close this chapter by sorting these out on a chart of the electromagnetic spectrum, illustrated in Figure 7-10, and defining some of the terms used to describe them.

The long-wavelength end of visible light is the *red* end. Beyond this lies *infrared*, which we normally detect through its heating effects. This fades grad-

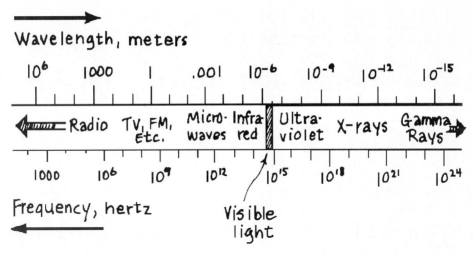

Wavelength, meters

FIGURE 7-10. The electromagnetic spectrum.

ually into *microwaves* and *radio,* with no clear boundaries. Today, our communications exploit all wavelengths from around 1 μm in the near infrared, which are used in fiber-optic systems, to ELF (extremely low frequency) signals with wavelengths of thousands of kilometers, which are used to contact submerged submarines. At the short-wavelength end, ultraviolet gradually fades into x-rays, with wavelengths comparable to atomic dimensions, and then to gamma rays, which extend out to wavelengths far smaller than an atomic nucleus.

All these forms of radiation except light were discovered since the days of Maxwell. His unified theory of electricity, magnetism, and light led directly to the discovery of some of them, and enabled us to understand the rest. There is probably no more outstanding example of pure science leading to practical results of overwhelming importance. So when scientists seek to justify spending the taxpayers' money on matters of no obvious or immediate practical importance, they thank the shade of James Clerk Maxwell.

Summary

Waves are patterns, usually distortions of a medium such as a string or a fluid, that travel out from their point of origin. In most instances they move at a constant speed, and several waves can be present in the same place at the same time. The terms *amplitude, frequency,* and *wavelength* are used to describe repeated identical waves. Two important wave phenomena are *standing waves,* which arise when waves are confined rather than allowed to spread freely, and *interference,* when waves from different sources arrive at the same place. Only certain specific patterns of standing waves may persist, and interference sets up patterns in which waves cancel in some places and reinforce in others.

CHAPTER 8

Does the Earth Really Move?

It may be that it does not move,
*   Or moves but for some other reason;*
Then let it be your boast to prove
*   (Though some may think it out of season*
And worthy of a fossil Druid)
That there is no Electric Fluid.

                    —JAMES CLERK MAXWELL

The experiment that opened the first rift in the seamless garment of New-tonian physics was performed in 1887 by a young professor at the Case School of Applied Sciences in Cleveland, Ohio. Like so many who have con-tributed to the building of America, Albert A. Michelson was an immigrant. Born in 1852 in the town of Strzelno in the German section of partitioned Poland, he was brought to the United States as an infant by parents who were fleeing the wave of intolerance and repression that followed in the wake of the failed revolutions of 1848. His father was an itinerant Jewish merchant, and his mother was Polish, a most uncommon match for that part of the world at that time.

Rather than settle into the teeming cities of the East and Midwest, Michel-son's father had the daring to join the California gold rush—not as a miner, but as a merchant catering to the miner's needs at Murphy's Camp in the Sierra foothills. Within the walls of their home the Michelsons provided some of the cultural amenities that the rough boom town so conspicuously lacked. To ensure him a proper education, Albert was sent to high school in San Fran-cisco, where he boarded with the principal. Recognizing Albert's aptitude for science, his host made him an assistant in the physics and chemistry laborato-ries. Blessed with good looks, a nimble mind, and skills that ranged from box-ing to the violin, Albert was a self-confident young man with seemingly bril-liant prospects.

But before Albert graduated high school the lode at Murphy's Camp played out, and his family followed the silver rush to Virginia City, Nevada. Short of capital to reestablish their business, the Michelsons simply could not afford to send a son to college. One way around this problem was to attend one of the service academies. Albert took the competitive examination for Nevada's lone appointment to the U.S. Naval Academy at Annapolis, but tied for first with a young man who was given preference as the son of a disabled Civil War veteran.

Undaunted, Albert made the long journey east, for Annapolis was changing. In the new Navy of iron steamships and long-range gunnery, a naval officer would have to be something more than a good sea dog, and a new curriculum had been adopted with hefty doses of science and mathematics. A few of the new plebes weren't up to the challenge, and withdrew in a matter of days. Michelson's name cleared the waiting list.

Albert graduated with high marks in everything but seamanship, so it seemed logical to keep him on as an instructor at the academy for his first tour of duty. He had married brilliantly, the daughter of a New York stockbroker. Soon he advanced his prospects even more with an experiment that defined his career, a precise measurement of the speed of light.

Michelson improved on a method illustrated in Figure 8-1, which had been developed by Jean Foucault, then France's leading experimental physicist. A beam of light was reflected off one face of a rotating octagonal mirror, and sent on a long path to a stationary mirror that reflected it back. On its return it reflected off another face of the mirror to a viewing eyepiece. In the microsecond or so of the round trip, the mirror rotated almost imperceptibly, but enough to shift the image a measurable amount. The known rate of rota-

FIGURE 8-1. Foucault's method for measuring the speed of light.

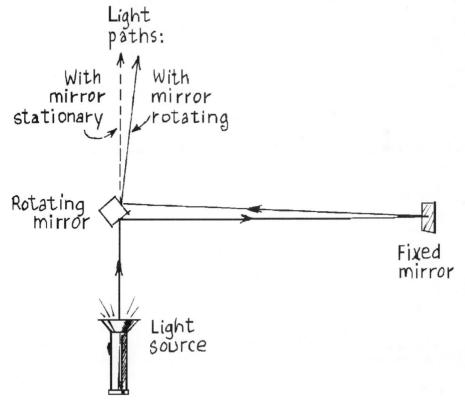

tion of the mirror was the "clock" that timed the light beam, and the result was good to about 1 percent.

Foucault had worked indoors, which limited the round trip to less than 100 feet. With a flair for the grandiose that would characterize his style, Michelson took the experiment outdoors to allow a half-mile path. The technical difficulties were enormous, but Michelson startled the world by coming up with a measurement more than twenty times more precise than Foucault's. This was front-page news in the *New York Times,* and the young man was ready to make his move in science.

Taking leave from the Navy, Michelson headed for Europe in 1879. For all the ingenuity of her inventors, America was still a backwater in basic science, so it was imperative to get at least some training abroad. His father-in-law offered to cover a year's living expenses, and telephone tycoon Alexander Graham Bell agreed to fund the construction of the prototype of an instrument that was to make Michelson famous. Bell operated an informal program designed to encourage young Americans to choose science over more practical careers. He was tickled by the young naval officer's audacity, for what Michelson had proposed was nothing less than measuring the absolute velocity of the Earth!

CLOCKING THE EARTH

A nineteenth century physicist did not need to be a diehard mechanist to believe in the aether. It is terribly hard to imagine a wave without some kind of medium to transport it. Sound waves have air, ocean waves propagate on water, and so on. Thus from the start, the aether was so intimately associated with the wave theory of light that few scientists could separate the two in their minds; to challenge the aether was to deny the wave nature of light.

In order not to impede the Earth in its orbit around the Sun, Maxwell's aether would have to flow either around or through the Earth without friction. As a result, light on Earth would move at different speeds in different directions. To see why, consider the movement of sound on a windy day. With a wind from the west at a steady 20 m/s, sound waves moving to the east will get a boost and travel at 340 + 20 = 360 m/s. Conversely, if headed west they will be slowed to 320 m/s. Only if we drift with the wind in a balloon will we find sound waves moving at the same speed in all directions. With the aether moving with respect to the Earth, a similar effect would be expected for light.

Even without an aether, common sense seemed to indicate that the wave theory of light should provide a standard of absolute rest or motion. If light waves travel at the same constant speed in all directions, surely they can only do so in one reference frame. In a frame that is moving with respect to that one, the wave travels at different speeds in different directions. There is one *privileged frame* that is different from all other inertial frames, the only one in which light moves the same speed in every direction. So although we will explain Michelson's experiment in terms of the aether theory, as he did, his analysis does not require the actual existence of the aether.

Still, in the minds of most physicists of Michelson's era, three concepts were indissolubly linked, as illustrated in Figure 8-2. These were:

1. The wave nature of light
2. The existence of a privileged frame
3. The existence of a physical aether

The existence of light waves seemingly implied a medium for them to move in, as well as a privileged frame in which they moved at the same speed in all directions. And if there is such a privileged frame, what has nature provided that makes it different from all other frames? An aether which is at rest in that frame seemed a logical answer.

As fast as light travels, the Earth's orbital motion is rapid enough to have a small but measurable effect on its speed. The Earth circles the Sun at about 30 km/s, one ten-thousandth the speed of light. And Michelson had already measured the speed of light to nearly a part in ten thousand. Unfortunately, however, a direct assault on this problem would not do. Michelson's measurement, like every other, had timed light over a *round-trip* course. If light were speeded up in one direction, it would be equally retarded in the opposite direction.

But Michelson had stumbled on a better idea, based on two significant insights: (1) the effect of the Earth's motion on the speed of light *does not cancel exactly* on a round-trip path, and (2) the tiny dimensions of light waves allow light itself to serve as a "ruler" for measurements of incomparable precision.

The effect for a round-trip would be 10,000 times smaller than on a one-way trip, so it would be necessary to measure to a part in 100 *million*. Michelson realized that while he could never *measure* the speed of light that precisely, with the aid of interference effects he could *compare* the speeds of light over *two different paths* to an even greater accuracy.

FIGURE 8-2.

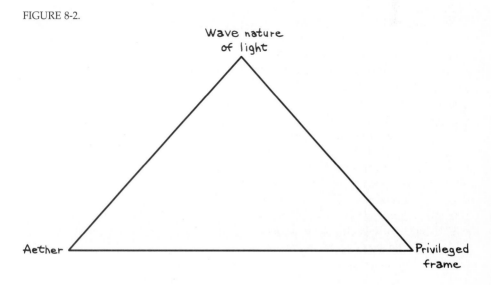

SWIMMING ACROSS A STREAM

Michelson's experiment rested on the realization that *it takes less time to swim across a stream and back than to swim the same distance upstream and back.* The mathematics required to understand this effect is not important in itself, but it does give rise to one key mathematical expression that is the basis for the quantitative aspects of Einstein's theory of relativity, so it is worth the trouble to go through it here. The result is more significant than the argument that leads to it, so the algebra that follows is not as important as the discussion that follows the formula.

Let a swimmer set out to cross a stream, headed straight for a point on the shore directly opposite, swimming at speed c in a current of speed v. To compensate for the current, the swimmer must point somewhat upstream. To someone standing on the shore the swimmer may appear to head directly across the stream, but to someone in a boat drifting with the current the swimmer follows a diagonal path, the hypotenuse of the triangle in Figure 8-3.

To estimate how much extra distance the swimmer must cover, we compare the hypotenuse of the triangle to its cross-stream side. The hypotenuse represents the swimmer's path in the water, at speed c. The downstream side of the triangle is the motion of the current, at v. By the theorem of Pythagoras, the squares of the two sides should add up to the square of the hypotenuse. We know the hypotenuse and one side, so we subtract and take the square root to find that the swimmer's speed, as seen from shore, must be $\sqrt{c^2 - v^2}$. The ratio of the distance covered in the water to the width of the stream is the ratio of the hypotenuse to this side.

FIGURE 8-3.

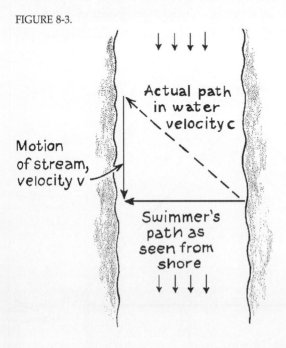

Actual path
in water
velocity c

Motion
of stream,
velocity v

Swimmer's
path as
seen from
shore

In relativity, this ratio is known as the *Lorentz factor* and is usually represented by the lowercase Greek letter gamma (γ). Completing the calculation, we get

$$\gamma = \frac{c}{\sqrt{c^2 - v^2}} = \frac{1}{\sqrt{1 - \left(\dfrac{v}{c}\right)^2}}$$

Because of its importance, the way γ varies with velocity is shown in the graph in Figure 8-4.

FIGURE 8-4.

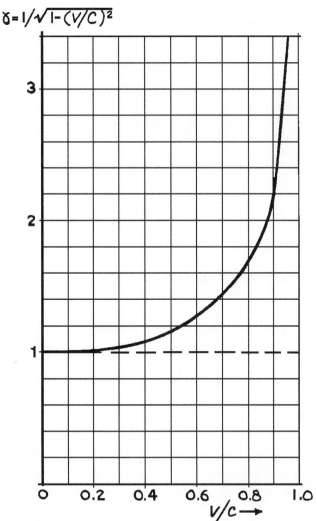

As a numerical example, suppose the swimmer can do 5 mph in a stream that moves at 3 mph. Then $\dfrac{v}{c} = 0.6$, its square is 0.36, and $1 - \left(\dfrac{v}{c}\right)^2 = 0.64$; since the square root of 0.64 is 0.8, we have $\gamma = 1.25$, which means that the distance the swimmer travels with respect to the water is 25 percent more than the width of the stream.

If $v = 0$, that is, the stream is stationary, then $\gamma = 1$ as we might expect, for the distance swum is exactly the width of the stream. This is as low as γ gets. It is always 1.0 or greater, which means that if you are trying to cross a stream in a perpendicular direction, the current is always a hindrance rather than a help.

Of far greater significance is the fact that the formula cannot be used at all if v is greater than c, for we would then have to take the square root of a negative number. In this situation, the stream moves faster than the swimmer, so there is no way to cross it without being swept downstream. In the theory of relativity, this restriction takes on a far more profound meaning.

The trip up and down the stream is somewhat more complicated, so we shall not derive the result. The swimmer moves upstream with velocity $c - v$ and returns with velocity $c + v$. But more time is spent on the upstream trip, so the average speed for the roundtrip is less than c. The ratio of average velocities is in fact γ^2, as readers adept in algebra are invited to verify for themselves.

For speeds much less than the speed of light, there is a simpler approximate formula for the Lorentz factor:

$$\gamma \simeq 1 + \frac{1}{2}\left(\frac{v}{c}\right)^2$$

This formula should be used only if v is less than about one-hundredth the speed of light.

A RACECOURSE FOR LIGHT

Now we turn to Michelson's apparatus. The reader has probably anticipated the next step. Replace the stream with the aether, and the swimmer with a ray of light, which is why we chose the symbol c, the conventional symbol for the velocity of light. But in this case, v is expected to be only about $0.0001c$, and γ thus differs from 1.0 by half a part in 100 million! How can we measure such a tiny difference?

The answer is that light carries its own yardstick, and a remarkably fine one it is. The marks on this yardstick are light waves—typically, about 0.5 µm long. What Michelson built was a device that set up a race between two light rays, with a way to judge the winner to a fraction of a light wave.

The racecourse, known as a *Michelson interferometer*, is depicted in Figure 8-5. A partially silvered mirror transmits half of the light falling on it, reflecting the rest, thus setting up two light beams at right angles. Two ordinary mirrors send these back along their paths, returning to the half-silvered mirror,

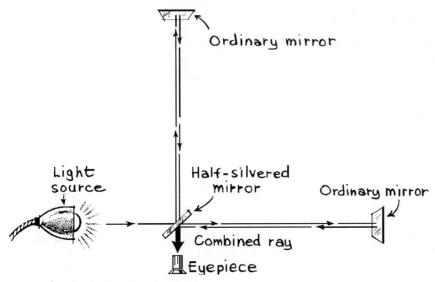

FIGURE 8-5. Michelson interferometer.

which recombines half the light in each beam. If the waves arrive crest to crest, the center of the pattern in the eyepiece will be bright. If they arrive trough to crest, it will be dark.

To make matters easier, he tilted one of the mirrors imperceptibly, so light in different parts of one beam would travel slightly different distances. The result was a pattern of parallel light and dark bars, like the ones observed by Young. It is easier to judge a change in such a pattern than a change in the brightness of light. If the relative length of the two arms of the interferometer changes by one wavelength, the whole pattern will shift to the left or right by a distance equal to the spacing between two bright bands.

Michelson then set the interferometer in a very slow rotation. If the aether were moving horizontally through the laboratory*, each light path would get a turn heading up and down stream and then, after 90 degrees of rotation, would become the cross-stream leg. The difference in length between the two paths—the key to the interference pattern—would change continually as the interferometer rotated, and the pattern should shift from side to side.

Building a successful interferometer was a technical challenge. It must rotate without changing its dimensions by even a small fraction of a wavelength of light. For his first attempt, in Berlin in 1880, Michelson tried using a structure of precision-machined steel. It was so sensitive to vibrations that when an assistant outside the laboratory stamped on the ground the effect was visible. Michaelson saw no shift in the pattern, but his first interferometer had been a trial run, and the expected shift could have been too small to see. Sim-

*Measurements were made around midnight when the direction of the Earth's orbital motion is due east.

ply getting the device to work had left a strong impression on the European professors to whom he gave demonstrations.

And in America in those years, a scientific name recognized in Europe was a saleable commodity. Michaelson applied for and won a position at the Case Institute, resigning his Navy commission. There he teamed up with Edward Morley, an astronomer who taught chemistry at the nearby Western Reserve University, which had better laboratory facilities than Case.

In the Cleveland version, the mirrors were mounted on a huge sandstone slab floated on a pool of mercury, with multiple mirrors to send each beam back and forth several times, lengthening the racecourse to 10 meters. But that was 20 million wavelengths of light, so a change of a part in 100 million would cause a shift of one-fifth of a band spacing. Michelson was sure he could detect a shift of one-hundredth of a band spacing, so this time there would be no doubt.

But there was also no shift! Michelson was bitterly disappointed, but did not let that slow him down. Leaving a puzzle for others to explain, he went on to a series of triumphant displays of experimental virtuosity, nearly all of which worked on variants of the interferometer idea.

The full story of attempts to explain away the Michelson-Morley experiment is interesting and at times amusing, for it still ranks as one of the biggest surprises in the history of physics. Hardly anyone was willing to abandon the aether theory on the basis of this one puzzling result. Most felt that Michelson had revealed a new feature of the aether, one that would help nail down some of the unknown properties of that mysterious substance. One attempt to follow this line of reasoning is worth mentioning, for it was the one that was on the right track, at least mathematically.

Following a suggestion by the Irish astronomer C. F. FitzGerald, the Dutch theoretical physicist H. A. Lorentz thought the key to the puzzle might be the electrical structure of atoms, which was discovered a few years after the Michelson result. If matter is held together by electrical forces that are transmitted by the aether, it might be that moving through aether wind "flattens" objects in the direction they are moving. It must flatten them by exactly the required amount, γ, to make the Michelson race come out in a dead heat. Of course, no material ruler could detect the effect, for it would conveniently shrink the same amount as whatever it was measuring.

PUBLISH AND PROSPER

This experiment and others gave Michelson's career a spectacular boost, for a new era had begun in American higher education, and gifted researchers were in short supply and great demand.

From colonial days, the model for American higher learning had been the English college. Teaching emphasized tutoring and student recitations, labor-intensive methods that left professors with little time for study or research. Serving a small and scattered population, colleges tended to be small, and professors had to be versatile enough to teach several subjects. Most colleges offered only a handful of strictly defined courses of study, with few or no electives.

But to the America of the late nineteenth century, growing rapidly in population and industrial might, another model of higher education was beckoning. The spectacular rise of the German economy, with particular emphasis on technically sophisticated industries like chemicals, steel, and electrical machinery, had impressed the world. Rightly or wrongly, the university system was given much of the credit for this growth.

At a German university, students launched on a specialized program of study from the first year, taught largely through lectures by august professors with plenty of time for personal research and the close supervision of candidates for the Ph.D. The faculty was organized into specialized departments, the strongest of which boasted well-established research institutes where a single professor dominated the work of a large team of scholars in junior positions.

After a protracted struggle between these two systems, America arrived at its own formula, which persists to this day. The German organizational structure was adopted, but democratized to preserve the more egalitarian collegiality of the English system. The lecture method became dominant, and in the larger colleges and universities the old general curriculum gave way to an elective system. Some liberal arts colleges preserve the vestiges of the English curriculum, though most have adopted the dominant system. The recent call for restoring a "core curriculum" is an echo of the English tradition.

In the last decades of the nineteenth century, one American university after another hopped on the bandwagon. The first were Yale and Johns Hopkins, followed by a number of large Eastern private universities. Among the public colleges, the Universities of Michigan, Wisconsin, and California followed soon after. Two new universities, Chicago and Stanford, were research universities from day one. Getting in early clearly paid off, for these schools remain among the leading research centers to this day.

Michelson played the game well, moving up from the Case Institute to Clark University, and then moving again to found the physics department at Chicago, with its generous libations of Rockefeller money. In his latter years, he moved one last time to the California Institute of Technology. Sticking to large optical instruments, Michelson had a number of celebrated successes. One of the most notable, for sheer physical size, was his last. To improve the measurement of the speed of light that launched his career, he built a mile-long vacuum pipe on the California desert. On his deathbed, he wrote the paper reporting the result.

Despite its growth, especially during and after World War II, American physics retains some of the Michelson flavor. Physicists in the English-speaking world tend to feel that experiment is the driving engine of progress, and many regard better instruments as the surest path to better experiments. Michelson is enshrined in the pantheon of great instrument builders. But the quality of his chisels was not what made Michelangelo a great sculptor, and experimental science, like sculpture, remains an art in which the creative imagination is as indispensable as it is intangible.

Be that as it may, building research instruments on a stupendous scale has long been a distinguishing feature of the American style. Many a physicist in the United States has carved an illustrious career out of skillful gadgeteering, without personally contributing much to physical thought. The giant particle accelerators, which are among the most expansive (and expensive) monuments ever erected to human curiosity, would probably have tickled Michelson's fancy.

Summary

An unsuccessful attempt by Albert A. Michelson to measure the "absolute" motion of the Earth, using the speed of light as a reference, had an important historical impact. For this measurement he invented an *interferometer*, a device that exploits interference effects to make measurements of exceptional precision. Michelson was the first United States Nobel laureate, and his career was closely linked to the rise of science and research universities in the U.S. The analysis of this experiment introduces the *Lorentz factor*, a mathematical function that will prove useful in Einstein's theory of relativity.

CHAPTER 9

The Birth of Relativity

But in physics I soon learned to scent out the paths that led to the depths, and to disregard everything else, all the many things that clutter up the mind, and divert it from the essential. The hitch in all this was, of course, the fact that one had to cram all this stuff into one's mind for the examination, whether one liked it or not.

—ALBERT EINSTEIN

Biographical sketches of famous scientists routinely claim that their brilliance was obvious from earliest childhood. Like many other stereotypes, this one does not fit Albert Einstein, by all odds the most celebrated scientist of the twentieth century, the only one whose face is recognized by nearly every educated person on Earth. But Einstein in his early years could only have been described as a "serious underachiever."

Einstein was born in 1879 in reasonably comfortable circumstances in Ulm, in the south of Germany, and raised in Munich. His family was in the electrical business, which in his youth represented the cutting edge of technology. Across the European continent, city after city was installing electric power. His uncle, an electrical engineer, and his father, a man perhaps a bit too gentle to be a real success in business, were partners in an effort to capture a small place in this growing market.

In his early years, young Albert alarmed his parents by being slow to talk, though when he finally got around to it, he spoke in complete sentences. He did seem alert and curious, and soon showed considerable talent for music. His uncle also noticed and cultivated Einstein's flair for mathematics. From the outset, he displayed an ability to willingly lose himself in deep thought.

Albert's happy childhood in the bosom of a warm and indulgent family took a turn for the worse when he entered high school at Munich's Luitpold Gymnasium. The watchwords at that institution were *discipline* and *authority*, two things that Einstein would detest throughout his life.

Though he did well in most subjects, Einstein's classroom demeanor infuriated his teachers. He would sit at the back of the room, dreamy-eyed and half smiling, denying them the respect and servility that they felt was their due. The headmaster was particularly offended. On the playground Einstein stood apart, averse to the rough-and-tumble of sports. At the end of his school day, he would retreat with relief to his home, his beloved violin, and what he later called his "holy geometry book."

Before the start of his final year, the Einstein brothers moved their failing business to the city of Pavia in Italy, leaving Albert behind to finish school. Deprived of emotional support, Einstein watched with apprehension the approach of his seventeenth birthday, a significant deadline for a young German of that era. After that date, he was liable for military service and could not leave the country without being regarded as a deserter. Not only would he be cut off from his family, but given his troubles with people in authority, he was sure that exchanging his school uniform for a military one could only mean disaster.

With an excuse provided by a friendly physician, Albert withdrew from school, renounced his German citizenship, and joined his parents in Italy. But their move had done little to improve the prospects for the family business. It became clear that Albert must train for some profession. He decided to take the entrance examination for the prestigious Swiss Federal Polytechnic School in Zurich, where he could study in his native German without having to return to Germany. His goal was modest enough—to become a teacher of physics and mathematics at the high school level.

In his chosen fields, Einstein's test scores were impressive, but he was terribly weak in foreign languages, chemistry, and biology. The director of the Polytechnic pointed out that he could bypass the examination if he could simply manage to graduate from a Swiss high school, which required a less demanding examination. He was steered to a school in the city of Aarau with a reputation for dealing kindly with free-spirited youth. It had been founded by educational reformers who stressed the visual imagination, which was Albert's strongest point. He later insisted that his best ideas always came to him in the form of visual images. The math and the words to explain them followed months or even years later. Aarau proved a happy and comfortable route to the Polytechnic.

Einstein's record in Zurich was decidedly uneven. He was disappointed that many of his courses were not up-to-date, so he cut lectures to study the latest textbooks on his own. He survived by cramming for the two batteries of exams that counted, with the help of careful lecture notes taken by his friend Marcel Grossmann. In the end, he led the class on the final exams for his teaching certificate. His school companions found him charming and witty, and a few even recognized his spark of brilliance. But by the time he graduated, he had antagonized several influential professors by his ill-concealed lack of respect. Furthermore, he was not yet a Swiss citizen. His job prospects seemed less than radiant.

For two years he struggled along as a substitute teacher and private tutor, until Marcel Grossmann made use of some family connections to land him the modest post of patent examiner in the Swiss capital of Bern. In this unlikely bureaucratic niche he was to enjoy some of the most productive years in the history of science. It provided him with a shelter from the pressures that normally beset a young scientist. In his later years Einstein was to comment:

Einstein during his Bern days.
(*Photo by Lotte Jacobi.*)

For an academic career puts a young man into a kind of embarrassing position by requiring him to produce scientific publications in impressive quantity—a seduction into superficiality which only strong characters are able to withstand. Most practical occupations, however, are of such a nature that a man of normal ability is able to accomplish what is expected of him. His day-to-day existence does not depend on any special illuminations. If he has deeper scientific interests he may plunge into his favorite problems in addition to doing his required work. He need not be oppressed by the fear that his efforts may lead to no results. I owed it to Marcel Grossmann that I was in such a fortunate position.

Life in Bern proved comfortable. His position actually paid better than the high school teaching posts he had aspired to, and he quickly acquired a small but lively circle of friends who met to drink and discuss deep ideas. In their youthful enthusiasm, they dubbed themselves the "Olympia Academy."

The fruits of this idyll were not just the theory of relativity, perhaps the greatest single-handed contribution to physics since the time of Newton, but some important steps toward the quantum theory. Einstein also played a significant role in persuading the last few skeptical scientists that atoms were real. Landmark papers on all these subjects appeared in one feverish eleven-month spree during 1905 and 1906.

Part of the Einstein legend holds that he had to labor for years in obscurity before the world finally recognized his achievements. In fact, his career after 1906 can be described only as meteoric. He was aided by the culture of central Europe, which set high value on the very qualities that set him apart.

Central Europeans of that era had seen too much social and political turmoil to put much faith in the enduring value of wealth, fame, or political power. They rested their cosmopolitan hopes for humanity not on established institutions, but on the power of the mind. The knowledge imparted by a good education was the one thing nobody could take away. Furthermore, they tended to believe that all human progress flowed from the work of a small number of geniuses. Once one was tagged as a potential genius, all doors were open and the rule book of career advancement went out the window.

Einstein was fortunate in that not only his friends but also a few influential scientists, especially Max Planck, professor of theoretical physics at the University of Berlin, ranked him as a genius. Planck formed that opinion simply by reading those 1905 papers, long before he met Einstein. Whatever their author might lack in credentials or position, these works displayed an exceptional clarity of thought. With his dreamy, intense gaze, Albert even *looked* the part of a genius. He rushed to completion his dissertation for the Ph.D., and aimed his sights higher than he had previously dared.

In the year 1909, Einstein had his "coming-out." Just turned 30, he was awarded an honorary degree, invited to address the leading figures in his field at a major scientific meeting, and offered an associate professorship at the University of Zurich. After two quick steps up the academic ladder, he was called in 1914 to a specially created chair at the University of Berlin, which granted him a generous salary and freedom from teaching duties any time he wished.

Einstein accepted this honor with mixed emotions. He was happy to be at the pinnacle of his profession, in daily contact with some of its best minds, for Germany was the world leader in scientific research, and Berlin was its most prestigious university. But he feared that he was now ". . . a goose that is expected to lay many golden eggs." He decided to hang on to his Swiss citizenship.

On his way to Berlin, Einstein stopped off in his boyhood home and paid a visit to his old school. The headmaster greeted him with reserve, convinced that this unpromising student must by now be destitute and had probably come to beg for money.

THE POSTULATE OF RELATIVITY

The central postulate of Einstein's theory of relativity is deceptively simple:

> The velocity of light is the same for all observers, in all directions, regardless of the motion of either the observer or the light source.

In terms of the "privileged reference frame" discussed in the preceding chapter, Einstein is saying *all frames are equally privileged.* How could this be so?

Obviously, it would take care of the Michelson-Morley result by *fiat.* One simply chooses a reference frame in which the interferometer is at rest. But the paradox is inescapable: If an observer finds a beam of light moving at c, how can another observer moving in the direction of the signal expect to measure its speed and get the same answer?

Though Einstein's step was a bold one, it was not totally out of harmony with the thinking of others. At nearly the same time, Henri Poincaré had come to a similar conclusion. Poincaré interpreted the failure of Michelson's and other attempts to measure the Earth's motion as evidence for some sort of general principle of relativity that would forever forbid the detection of absolute motion. But Poincaré was nearing the end of a long and distinguished career, and was hoping to solve this problem within the context of the aether theory. It took the nimbler mind of a younger man, experienced enough to understand the problem but not yet committed to conventional answers, to dare the final step—to take this principle as the *starting point* of his argument, rather than a conclusion to be worked up to.

Einstein's insight was to see that while the apparent problem was the motion of light, the questions it raised went far deeper than that. He was ready to reexamine concepts as basic and seemingly unalterable as *space* and *time.*

The arguments by which Einstein developed relativity are not mathematically difficult, but in order to follow them one must engage in a kind of thinking that is very unusual. One must *follow logic even when it seems to contradict common sense.* Most people—even most scientists—rarely think in that fashion. Good thinking usually moves forward on two legs, with logic and intuition both playing their part. Because of the unfamiliarity of this mode of thinking, as much as the strangeness of the conclusions it leads to, it is wise to remember that you *should* find them baffling, at least on first acquaintance.

HOW TO USE THE POSTULATE OF RELATIVITY

Einstein developed the theory of relativity largely by visualizing situations that he called *gedanken experimenten*, or "thought experiments," because for practical reasons they could not actually be performed. This has become one accepted way to teach the subject, and we will adopt it here. But it helps to lay down some guidelines to steer you through the thickets of logic to come:

1. You may always assume *you* are at rest, and light travels at *c* with respect to you.
2. Observers you regard as moving are equally free to assume they are at rest and to construct their own pictures of reality.
3. Nonetheless, all observers must *accept all observations*, their own or another's, as valid. If we can imagine a situation that leads, through the application of the postulate, to an unavoidable dispute over *observations*, then either the postulate of relativity is false or *that situation may not occur in nature!*
4. The disagreements will concern things *inferred* from observations. Relativity will teach us that certain things we instinctively take to be observable features of the world are really constructs of the human mind.
5. All disputes arise from *one kind* of inference: *estimating when a remote event happened*, by calculating how long it took the news to reach you at the speed of light.

SHIPS THAT PASS IN THE NIGHT

Two of the most direct consequences of Einstein's postulate are that the speed of light is the upper limit for all velocities, and that observers moving with respect to one another may not agree that two events happened at the same time. Our first *gedanken* experiment will demonstrate these points.

Imagine two spaceships passing each other in the far reaches of outer space with a relative velocity just below the speed of light. As they meet, let a bright flashbulb go off between their ships, as illustrated in Figure 9-1. Rule 1 allows the crew of each ship to believe that it sits serenely at the center of a sphere of light, which is expanding in all directions at velocity *c*. The other ship is, of course, somewhere else. A sphere can, after all, have only one center. Still, rule 2 obliges us to find a way for both of these pictures to be right.

Let us allow the two crews to turn around and have another go at it, to resolve the disagreement by observation. At this point, rule 4 plays a crucial role, for it reminds us that the sphere is *not something we can observe!* Light is moving in all directions, and we can see it at only one place and time. We must translate the statement "I am in the center of a sphere" into something that can be tested by observation. It then becomes *"at any instant, all the light heading out from the flash is the same distance from me, regardless of direction."*

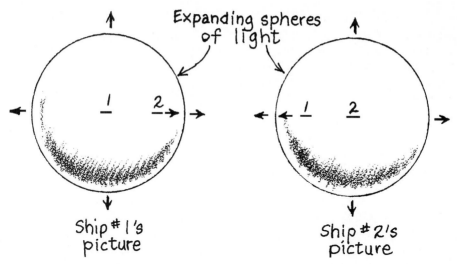

FIGURE 9-1. Two conflicting "pictures."

The simplest way to test this assertion is to reflect some of the light back. One crew agrees to rig reflectors far from their ship, at equal distances ahead and behind. If the flashes from both reflectors return to the ship at the same time, it must be at the center of the sphere. To prove their point they can even produce a photograph, taken by a camera with a fast shutter, showing both reflections at once. But this does not resolve the dispute, for reasons illustrated in Figure 9-2.

The other crew has an equally ready explanation of what happened. Since they see both ship and reflectors in rapid motion, the rear reflector rushed forward to meet the light headed toward it, while the reflector in front rushed away. Clearly, light must strike the rear reflector first, and the front reflector some time later. After reflection, the two light beams have reversed direction. The rear beam now must catch up with a fleeing spaceship, while the one from the front finds the ship rushing to meet it. As the illustration clearly shows, each round trip consists of a short and a long leg. For the light ray that heads to the rear, the short leg comes first, and the long leg later. For the ray toward the front, the order is reversed. But the lengths of the two legs are the same in both cases, so for both beams the round-trip times are the same, regardless of the order of the short and long legs.

So this crew concedes that the flashes from both reflectors *reach the ship* at the same time. What they deny is that they *hit the mirrors* at the same time! They concede the validity of the observation but deny that it proves their rival was in the center of the sphere.

So the dispute is reduced to the question of whether light hit the two mirrors at the same time, which *cannot be resolved by observation!* Each crew can cling securely to its own interpretation. Since no one can experience directly what is happening in two different places, there is no way to settle the argu-

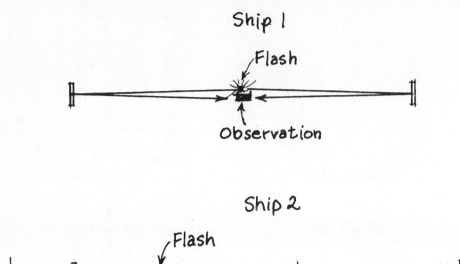

FIGURE 9-2.   Light paths according to two ships.

ment. They are both right. The postulate of relativity has withstood its first test. We must pay a price, however. We are obliged to discard the idea that the phrase "happened at the same time" has an unambiguous meaning that all observers must accept, as long as two things happen in different places. This seemingly innocent assertion has been removed from the realm of *fact* and consigned to the realm of *convention*.

AN UNBREAKABLE SPEED LIMIT

Imagine, in the preceding example, that the relative velocity of the two ships was *greater* than *c*. In that event, each crew would insist that the other is actually *outside* the expanding sphere. Indeed, they could not possibly have even seen the original flash! But rules 1 and 2 assure us that both crews must always see the flash. Since rule 3 allows no disagreements on actual observations, speeds greater than *c* must be ruled out.

This is the relativistic significance of the rule that *v* must be less than *c* to compute γ in the preceding chapter. Velocities greater than *c* are simply illegal. Of course, rule 3 allows us another option—to abandon the postulate of relativity—but given our lack of experience with things moving this fast, nothing obliges us to do so.

Put another way, if something moves at a speed that is the same in all reference frames, there is only one such speed and it is the limit of all possible speeds. There cannot be two different speeds both of which obey the Einstein postulate. This argument drives home a subtle but crucial point: the postulate of relativity is not a statement about *light itself,* but simply a statement about the *speed* of light!

This represents a drastic change in our attitude toward the speed of light. No longer is it an accidental property of light or of the medium in which light travels, for the logic of this argument makes no reference to the nature of light or to any medium. The *only* thing we have assumed about light is that it has the same speed in all reference frames. Thus we must have stumbled upon something more universal. We shall see in the next chapter that the true significance of the speed of light is not that it is an accidental property of electromagnetic waves, but that it is a fundamental scale factor in the universe between *space* and *time.*

At this point we are in no position to say how this speed limit is enforced. Why can't a rocket simply continue to accelerate until it exceeds the speed of light? In Chapter 11, when we come to grips with the changes in Newton's laws required by relativity, we shall find the speed limit is self-enforcing.

THE ELASTIC TRAIN

We move now to our second *gedanken* experiment, one of Einstein's own favorites, measuring the length of a moving object, which he dreamed up while riding to work at the Patent Office through the narrow streets of Bern. Of course, in his mind's eye he replaced his lumbering streetcar with a fast express moving close to the speed of light.

It is important to be perfectly clear that a moving object means one that is moving *with respect to the ruler* that is being used to measure its length. If the ruler is moving with the object, then you are measuring the object's length in a reference frame in which it is *not* moving.

Let us imagine that alongside the track is a string of telegraph poles, spaced at a known interval, with each pole bearing a clearly visible number. This is our ruler. To measure the length of the train against this ruler, simply note which pole is at the front of the train, and which at the rear, *at the same instant.* After our first *gedanken* experiment, we realize that this requirement is not as innocent as it sounds.

One way of doing this is illustrated in Figure 9-3. At the center of the train, place a light visible from both ends. When they see the light flash, conductors at each end of the train note the number of the nearest pole. But a signalman on the ground immediately yells "foul"—the measurement is invalid, because the poles were not observed at the same time!

He sees it as illustrated in the bottom half of Figure 9-3. The conductor at the rear of the train was moving toward the flash of light and saw it before his colleague at the front, who was moving away and made his observation later.

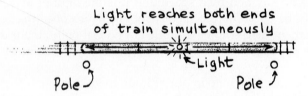

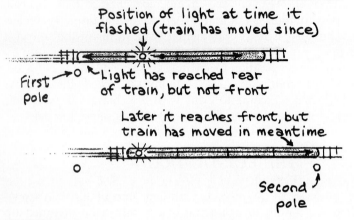

FIGURE 9-3.

In the meantime, the train moved, so the poles are farther apart than the length of the train. The signalman believes the train is *shorter* than the distance between the poles noted by the conductors.

There is a logically consistent way to resolve this dilemma: *moving objects must shrink along their line of motion.* Then both the train crew and the observer on the ground can agree that *the distance between the poles, in the ground frame of reference, is greater than the length of the train.*

The train crew believes this is so because the "ruler" has shrunk—the poles are closer together, so more can be fit in between the ends of the train. The signalman explains that the distance between the poles is greater than the length of the train measured at rest because it is not a valid measurement of the train alone. Instead, it is the sum of *two* distances—the length of the train *plus* the distance that it moved between the observation at the rear and the one at the front. Thus the length of the train could actually be *shorter* than when it is at rest, and still have the poles farther apart than the rest length of the train.

There is no quarrel about observations. The signalman accepts that each conductor noted the number of the nearest pole at the moment he saw the flash. What he disputes is that both did so at the same time. That is an *infer-*

ence—there is no way to compare times in two places by direct observation. It is not part of the sensory experience of any participant, nor can it come from reading a single instrument. Instead, it is part of the picture of reality we make *in our minds.* Einstein has redefined the boundary between what is "out there" and what is in our heads. Nature doesn't make reference frames—*we* do.

Though the choice of a reference frame is arbitrary, Einstein assures us that it is still valid and useful, as long as one takes into account some of his peculiar effects. Thus two signalmen could stand a distance apart that is shorter than the length of the train at rest, and safely cross the tracks at what *they* regard as the same time, with the train safely between them.

This was Einstein's analysis of the Lorentz contraction, and we shall see in the next chapter that it is quantitatively the same. But where Lorentz viewed it as a mechanical effect based on motion with respect to the aether, Einstein saw it as a consequence of the choice of reference frame. For the signalman, the train has shrunk, while its crew see no such effect. For Lorentz, the v in his formula is the velocity with respect to the aether. For Einstein, it is the velocity of an object in some arbitrarily chosen reference frame. Why not make life easy for yourself, and choose one in which it is zero?

In the light of this, let us reconsider the Michelson-Morley experiment. In the Earth frame of reference, there is no shrinkage and no effect, since light moves at the same speed in both arms of the interferometer. In a frame in which the Earth is moving, the shrinkage assures that both beams travel the same distance, as Lorentz described. Both reference frames give the same observed result. There is no privileged frame, which makes it hard to go on believing in the aether.

All of this flies in the face of common sense only because we intuitively assume that what we are *seeing* now is *happening* now. This is perfectly reasonable; all of our experience tells us that nothing moves or changes much in the microseconds or less that it takes for light to reach us from objects nearby. But when dealing with fast-moving objects, these brief delays become very important.

We can exploit Einstein's train to hint at yet one more relativistic effect. Suppose the conductors, in order to avoid problems with light signals, meet at the center of the train to synchronize watches, and then take up their stations at the ends and observe the poles at a prearranged time. In a self-consistent world, this must give the same result as the previous method. The signalman must still deny that the poles were observed at the same time. So something must happen to these watches in the time between when they were set and when the observation was made. We will see just what it is in the next chapter.

THE GARAGE PARADOX

We close this chapter with a third *gedanken* experiment. Imagine a garage with doors at both ends, set to open automatically when a car approaches and close when it clears the door. Imagine also a car that at rest would be as long as the garage. The car sails through the garage at a speed approaching that of light; it is thus shortened, at least from the point of view of someone in the garage. The rear door opens to admit the car, and then closes behind it before the front door opens to let it out. For an instant, the car is inside a closed garage.

But from the point of view of the car's rather reckless driver, it is the garage that gets shortened. As he sees the process, the car must at some point stick out at both ends, and both doors must have been open at once! These two conflicting pictures are illustrated in Figure 9-4. Common sense tells us one or the other must be wrong.

Again, relativity insists that this is not a question with an unambiguous answer. The secret lies in the difference in the sequence of door openings and closings, as seen by the two observers. These are summarized below:

Garage Frame:	Car Frame:
rear door opens	rear door opens
rear door closes	front door opens
front door opens	rear door closes
front door closes	front door closes

Note the reversal of the order of the second and third events. The question of whether or not a garage is "closed" turns out to be yet another example of comparing times in different places. In the next chapter, we will examine what the driver and an observer in the garage really *see*, in order to develop this example more fully.

It cannot be emphasized too strongly that relativity does *not* deny that there exists a single reality, and admit only the multiple, conflicting personal realities of different observers. Einstein insisted that science describes things that exist independent of any observer. In this example, the car is real, the garage is real, all the openings and closings of doors are real, and so is everything the driver and others see. It is the *pictures* shown in the illustration that are, in a sense, not "real." They are conventional ways of representing reality. In the next chapter we will show that these pictures do not represent what *either* observer *actually sees!*

The word that best describes how one obtains such a picture is the verb *construct*. The picture does not represent immediate experience. Instead, it is constructed from observations by correcting for the time it takes for light to reach the observer. It takes the quantitative apparatus developed in the next chapter to do this properly.

There is nothing new or even particularly "scientific" about this procedure. Artists may portray three-dimensional reality on a two-dimensional canvas by the use of one convention, that of *perspective*. Engineers, on the other hand, tend to prefer the three-view mechanical drawing. Nature does not tell

A car passes through a garage at six-tenths the velocity of light. The car is actually slightly longer than the garage

An observer in the garage ~~feels~~ there is an instant when the shrunken car is inside a closed garage

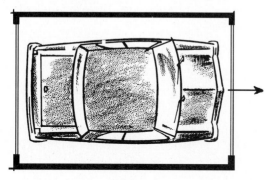

...while the driver of the car, who believes the garage has shrunk, feels quite differently

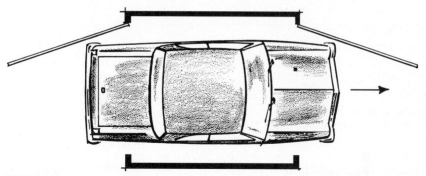

FIGURE 9-4.

us which to choose. Neither can depict the totality of an object, any more than a balance sheet tells you all there is to know about a business.

If the books don't properly or conveniently reflect the state of a business, a new system of bookkeeping can be adopted. In the next chapter, we will see that this is also true in physics. The "better bookkeeping" system is called *four-dimensional space-time*.

Summary

Albert Einstein's temperament, upbringing, and education prepared him to think in novel ways, but did not seem to point to a brilliant career until 1905, when he produced five papers of exceptional quality. Two of these outlined what came to be known as the *theory of relativity*. Its starting point was the postulate that light will travel at the same speed in all directions in all reference frames. He then accepted modifications to our notions of space and time that were required to make this possible, stipulating only that they must not conflict with observations. Through a series of "thought experiments" *(gedanken experimenten)* we demonstrate that the postulate rules out motion faster than the speed of light, and that it leads to disagreements over whether two events in different places happen at the same time. These in turn lead to disagreement as to the length of a moving object.

CHAPTER 10

The Wedding of Space and Time

Alice laughed: "There's no use trying," she said; "one cannot believe impossible things."

"I daresay you haven't had much practice," said the Queen. "When I was your age, I always did it for half-an-hour a day. Why, sometimes I've believed in as many as six impossible things before breakfast."

—LEWIS CARROLL, Through the Looking Glass

There are two ways to resolve a disagreement. One is to *learn to live with it*, as long as all parties thoroughly understand and tolerate each other's points of view. The other is to *find a common ground* on which all parties can agree. The theory of relativity offers both of these ways to resolve the disputes between observers implied by Einstein's postulate.

The "live with it" approach was outlined in Einstein's first paper on relativity, submitted to the journal *Annalen der Physik* in June 1905. It allows any observer to translate the picture of reality in any reference frame to that in any other. The task of this chapter is first to show how to carry out these translations.* This exercise will reveal that our usual way of describing nature is actually more formal and artificial than we realize. This will prepare the way for the "common ground," the four-dimensional *space-time* convention.

We found in Chapter 9 that whenever we use information gathered in one reference frame to construct a picture in another frame, we must accept three adjustments that seriously conflict with our commonsense notions of space and time:

1. Moving clocks appear to run slow.
2. Moving objects appear shortened along their line of motion.
3. Events that are simultaneous in one reference frame may not be in another frame.

To reduce the confusion, it is essential to deal with these effects one at a time. We will treat them in the order stated above.

---

*In the standard terminology of relativity, they are called *Lorentz transformations*.

HOW SLOW DOES THE CLOCK RUN?

The postulate of relativity requires that any measurement of the speed of light, regardless of the frame of reference in which it is made or the direction in which the light is moving, must give the same numerical result. To examine the implications of this postulate for measurements of time, we will consider a measurement of the speed of light made on a moving train.

The result must show that light moves at the expected speed *with respect to the train.* But an observer on the ground should be able to use the *same measurements* to show that light moves at the very same speed with *respect to the ground.*

This measurement is made with a *single clock* located on the train, timing a light ray that makes a round-trip *across* the train, perpendicular to its motion. The single clock removes the problem of synchronizing clocks in different places. A path across the train is chosen because there is no argument about the width of the train, since there is no motion of the train with respect to the ground in that direction.

The observer on the ground is bound to accept the validity of all measurements made on the train, but is otherwise free to interpret them in whatever way it takes to show that light moves at the expected speed with respect to the *ground.*

It is clear from Figure 10-1 that the two observers disagree about the distance traveled by the light ray. While an observer on the train believes it has returned to its starting point, the one on the ground believes it has traveled forward and thus has gone farther. That is why no measurements are made in the ground frame, for that would require two clocks in different places. Geometrically, the situation is exactly the same as for the swimmer in Chapter 8. Thus if the train has width w, the observer on the train believes the light travels exactly $2w$, while the one on the ground believes the true distance is $2\gamma w$.

Now let each observer calculate the speed of light. The postulate of relativity insists that both should get the same answer, c, which is 300 million meters per second. The observer on the train divides by the time t measured on the clock:

$$\frac{2w}{t} = c$$

but to the observer on the ground, the numerator should be $2\gamma w$. How can the answer still be c? One can't tamper with w—both observers agree on the width of the train. There is only one way to make the answer come out right: multiply the *denominator t* by γ also! Then the gammas will cancel and the result will be the same:

$$\frac{2\gamma w}{\gamma t} = c$$

What is the significance of the γt in the denominator? Since the numerator represents the distance traveled by light *in the ground frame,* the denominator must be the *time elapsed* on a clock in that same frame, so that the speed of light

Path of light ray across train

As seen by observer on train

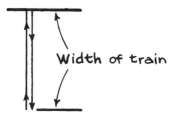

Width of train

As seen by observer on ground

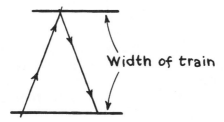

Width of train

FIGURE 10-1.

is its speed in that frame. Since multiplying by γ increases the value of t, it means that more time elapses in the ground frame than is recorded on the moving clock. This is the meaning of the statement *moving clocks run slow.*

But the observer on the train is free to consider any clock on the ground to be moving, so *it* is the one that runs slow. Each observer believes the other is using a slow clock.

If you balk at the notion that two observers can simultaneously believe that the other's clock is slow, keep one important fact in mind: two observers in relative motion get at most *one chance* to compare clocks face-to-face as they pass. Thereafter they are moving rapidly apart. They can communicate only by light or radio signals that take a noticeable time to reach the other party, and as they move farther and farther apart successive replies take longer and longer. Each estimates the other's clock reading by correcting for this time lag, assuming the other party is the one who is moving. The discrepancy in their corrections just continues to grow. A numerical example later in this chapter will show exactly how this works.

Returning to the train example in the preceding chapter, let the conductors meet at midtrain to synchronize their watches. The signalman will claim that by walking on the train in opposite directions, the conductors change their speeds with respect to the ground. The conductor who walks toward the front of the train moves faster, so his watch runs slower than a watch at rest with respect to the train. The other conductor experiences the opposite effect, so his watch runs faster. By the time they reach the ends of the train their watches are no longer synchronized. So once again, the pole opposite the rear is observed earlier than the one at the front.

THE FITZGERALD SHRINKAGE

Now that we have the quantitative measure of the apparent slowdown of a moving clock, we can use the result to derive the apparent shrinkage of a moving object along its direction of motion. Again we roll out Einstein's beloved train. One easy way to measure its length without ever having to be in two places at the same time is to measure its speed and then time its passage with a stopwatch.

An observer on the train would challenge the validity of that measurement—"You did it with a slow clock, so your value is too short." As in the measurement of the speed of light, he would multiply the time on the stopwatch by γ to get a value he believes. Since he then multiplies the time by the velocity to get the length of the train, his value for it is γ times longer than the estimate of an observer on the ground.

If relativity is self-consistent, this corrected measurement must agree with one made with a ruler that is at rest with respect to the train. Thus if the observer on the train makes such a measurement and reports it to the observer on the ground, that observer must *divide* by γ to get the "moving length" of the train. This is of course an odd way to measure the length of a train, but if relativity is valid, this way is as good as any other and will yield the same result.

To summarize these two effects, here are the "translation rules" for converting measurements in one reference frame to those in another:

If a moving observer reports a *time interval*, then *multiply* by γ to get the time interval in your frame.

If a moving observer reports a *length* along the direction of motion, then *divide* by γ to get the length in your frame.

If you prefer to see this expressed in formulas (if not, skip this paragraph and the equations that follow it), let L be the length of a moving object (along the line of motion), W be its width (perpendicular to the motion), and t be a time interval. Let the subscript zero stand for the quantity as measured in its rest frame (L_0 and W_0 measured by rulers at rest with respect to the object, t_0 in the rest frame of the clock). Then we have, as measured in a frame in which the object or clock are moving:

$$L = \frac{L_0}{\gamma}$$

$$W = W_0$$

$$t = \gamma t_0$$

If the reader is now thoroughly dazed (a common feeling at this stage in the study of relativity) by a world in which moving objects shrink and moving clocks lose time, a numerical reminder may prove reassuring. The most rapid large human artifacts (interplanetary rockets) move at speeds around one ten-thousandth of the speed of light. Thus, γ differs from one by about one part in 100 million. For more leisurely vehicles, such as supersonic jets, the effect is more like one part in 100 billion.

A CASE STUDY OF RELATIVISTIC TIME

This example, cast in the form of a brief dialogue, is designed to show how it is possible for two people in motion relative to one another to *each* believe that the *other's* clock is slow. The point to be made is that any comparison of the clocks involves communications over distances that are both *large* and *changing*. In order to understand what is going on, they must take into consideration the long time lags involved in communications over immense distances. We will show that if they ignore the relativistic corrections, they will disagree as to what they should observe, while if they perform them properly, they will agree on observations, though they disagree on other features of the process.

Two astronauts (Joe and Sue) pass one another in the far reaches of interstellar space, at a relative velocity of $0.6c$. At the instant they flash by one another, they synchronize watches, and Sue agrees to send a radio message (which travels at the speed of light) ten minutes later. The ensuing dialogue follows:

SUE: Ten minutes at the tone—ding!

JOE: Aha! Just like Einstein said, you have a slow clock! If you'd *really* called back in 10 minutes, you'd have been 6 light minutes away, and I would have gotten your message in 10 plus 6 equals 16 minutes. I must sadly inform you that my clock read 20 minutes at your ding. But relativistically, for your motion, gamma is 1.25. You really transmitted at 12.5 minutes, when you were 7.5 light minutes away. That adds up to 20, like I said.

SUE: Okay, knucklehead, reread your Einstein! He allows two to play at this game, and I can just as easily show that *your* clock is the slow one! I say *you're* moving, and were only 6 light minutes away when I sent. But you were running away from my message, which had to catch up at a relative speed of $0.4c$. It took 15 minutes to make up your 6 light minute lead, so you really got it at 25 minutes. You say 20, so *yours* is the slow clock.

JOE: I guess I gotta concede that you know what *your* clock said when you *sent*—I know what *mine* read when I *received*—and these are the *only* facts we have to work with. We're arguing about what my clock read when you sent, or yours read when I received, which has nothing to do with actual experience, since we're hundreds of millions of miles apart!

SUE: Right you are! Isn't it odd how we *must disagree* on these calculations, *in order to agree* on the observations!

Without the relativistic corrections, Joe would have expected his clock to read sixteen minutes when he got Sue's message, while Sue would have expected it to read twenty-five. With the corrections, both agree it should read twenty.

It must be emphasized that each is perfectly free to adopt the other's point of view, or to regard *both* spaceships as moving (that makes for a much more tedious calculation). The point is that relativity insists that *it makes no difference which reference frame you use*—all account equally well for the two observable facts.

What happens if one of the astronauts turns around and comes back for a direct face-to-face comparison of clocks? The answer is that it matters crucially *which* does the turning around. This is the famous "twin paradox," of which we shall hear more later.

The 2 to 1 ratio between time on the *sender's* clock and the time on the *receiver's* clock applies to all subsequent transmissions. Joe's reply, sent when his clock read twenty, arrived when Sue's read forty, and her reply reached him at eighty minutes. For $v = 0.8c$ the ratio would be 3 to 1.

The calculations are summarized in Table 10-1. For convenience, distances are measured in light-minutes (lmin), the distance light travels in a minute, which is more than 10 *million* miles! The speed of light is then 1.0, and the relative velocity of the ships is 0.6, measured in *light-minutes per minute*.

Note that of all the numbers in this table, *only two*—the ones in boldface—represent actual observations. The rest are part of the superstructure we erect to obtain a working picture of reality, one in which everything we experience is understandable.

BACK TO THE GARAGE

It is appropriate at this point to give substance to the "pictures of reality" referred to in rule 2. When dealing with rapidly moving objects, these pictures are *not what is seen*, which shows different parts of things at different distances and therefore at different times. In this *gedanken* experiment we will see how two observers, momentarily in the same place at the same time, *see essentially the same image*, but use it to construct *contradictory* pictures.

Assume the car and garage in Chapter 9 have a relative velocity of 0.6c, and let each be 20 feet long. The use of *feet* is justified by a simple relationship; the speed of light is about 1 foot per nanosecond (ns, a billionth of a second). Let our friend Sue be the driver, and have her seated halfway between the front and rear bumpers of the car. Standing midway in the garage is Joe.

TABLE 10-1

Quantity	Sue's Analysis	Sue's Value	Joe's Value	Joe's Analysis
What Sue's clock read when she transmitted	I know what I saw!	**10 min**	10 min	I accept Sue's observations (rule 4)
The "true" time when Sue transmitted	I trust my own clock (rule 1)	10 min	12.5 min	Moving clock, runs slow by $\gamma = 1.25$
How far apart the ships then were	Time × velocity $= 10 \times 0.6$	6 lmin	7.5 lmin	Time × velocity $= 12.5 \times 0.6$
Relative velocity of the message and Joe	He's moving, so $c - 0.6c$ (rule 2)	$0.4c$	c	I'm standing still (rule 1)
Time the message spent in transit	Distance ÷ velocity $= 6 \div 0.4$	15 min	7.5 min	Distance ÷ velocity $= 7.5 \div 1.0$
Total time since the ships met	Add the times: $(15 + 10)$ min	25 min	20 min	Add the times: $(12.5 + 7.5)$ min
What Joe's clock read when he received	Moving clock, runs slow by $\gamma = 1.25$	20 min	**20 min**	I know what I saw!

Let us focus in on the instant when Sue passes Joe, at the center of the garage. Since both observers are instantaneously in very nearly the same place, what they actually observe at this instant must be roughly the same.

If both are looking in the direction in which the car is moving, they will see that the *front bumper of the car is halfway to the door.** Each realizes this is old news—it took some time for the light to reach them. In the meantime, motion has taken place, and in order to decide where things are *now,* a correction must be made for this motion.

Joe is concerned with the front bumper of the car—has it moved far enough to reach the door? In Sue's reference frame it is the door that is moving, and she wants to know if it has reached the front bumper. The key to the matter is that each makes the correction differently, because they use different scales to measure distance.

Each uses a yardstick that he or she regards as at rest. For Joe, this is the *door,* which he knows is 10 feet away. Hence the car's front bumper must be 5 feet away. Sue chooses the *car* as reference, and since its front bumper is 10 feet from her, the door must be 20 feet away! Table 10-2 summarizes how each corrects the observed picture for the motion that took place while the light was on its way.

*This can be shown by working backward the calculations presented in Table 10-2.

Since the door is the farthest point of the garage, and the front bumper the farthest point of the car, Joe concludes the car is shrunken, while Sue is equally sure the garage is shrunken. The difference comes about because Joe is correcting for the motion of the bumper, which he sees as only 5 feet away, while Sue is correcting for the motion of the door, which she sees as 20 feet away. Thus her time correction is four times as big as Joe's, allowing for four times as much relative motion.

This is the very essence of relativity: *we never see remote objects as they are now,* only as they were some time ago. If they are moving, one must correct for this motion, and there is no one unique correct way to do this. The "slice in time"—an extended region of space all at the same instant in time—is a *fiction,* a construct of the human mind that does not correspond exactly to the way we actually experience reality.

Of course, in our everyday world, where things don't move so fast and nanoseconds are ridiculously short time delays, it is a pretty *good* fiction, so we intuitively think that way. Also, it is mathematically convenient to do so. What Einstein is saying is that we can go on using this fiction, but must recognize it for what it is, and allow that different observers may construct it differently.

One might ask "why *bother* with these corrections—isn't what we actually *see* a reasonable working picture of reality?" The answer can be gleaned from this example by asking what our two observers would *see* if they looked toward the *rear* of the car.

Here the situation will be reversed; the car is *protruding,* and the *door* is halfway along the rear half of the car. This is a general rule: objects that pass near you at close to the speed of light appear hideously distorted. The part that has gone by you appears *compressed,* while the part coming toward you looks *stretched.* This is because you see the farther parts as they were at an earlier time. If they are coming toward you, you see them farther away, and if they are headed away, you see them closer. To Joe, the rear half of the car looks four times as long as the front half! Sue sees the *garage* similarly distorted.

TABLE 10-2

Quantity	Joe's Value	Sue's Value
Distance to fixed reference point	10 ft (door)	10 ft (bumper)
Apparent distance to moving object	5 ft (bumper)	20 ft (door)
How long ago light left it	5 ns	20 ns
Motion since then (+ means *away*)	+0.6 ft/ns × 5 ns = +3 ft	−0.6 ft/ns × 20 ns = −12 ft
Distance to moving object now	5 ft + 3 ft = 8 ft	20 ft − 12 ft = 8 ft

Given these distortions, they would probably prefer to correct by the relativistic rules. What they *see* is even farther from an acceptable picture of reality in an orderly world. At least Einstein allows the car to retain a constant shape!

Einstein has, in effect, relocated the boundary between what is out there in nature, and what is constructed in our minds. To a previous generation of scientists, the pictures in Figure 9-4 and the corrected positions in Table 10-2 would have been considered just as real as observations. Einstein realized that they are true only by convention, in an arbitrarily chosen reference frame. Useful as reference frames may be, nature doesn't hand them to us—we make them up for ourselves.

We will soon see that there is yet another way to represent reality—the four-dimensional *space-time* picture. Though it is less intuitively satisfying than our normal pictures, it has the advantage that all observers can agree on a common view.

THE RELATIVITY OF SIMULTANEITY

As a final step in developing the relativistic translations, let us put in quantitative terms the gap in time in one reference frame between two events that are simultaneous in another frame. This comes about because each observer feels that the other has moved while the light signals that bring the news of the events were in transit. If one observer believes an event occurred a distance L away, the signal took the time $T = L/c$ to get there. During this time the other observer moved a distance vT. Thus their estimates of the time that the signal was in transit must differ by the time it takes light to travel this additional distance, which we get by dividing this distance by c:

$$t = \frac{vT}{c} = \frac{v}{c} \times \frac{L}{c} = \frac{Lv}{c^2}$$

Two events that are simultaneous in a frame moving with velocity v with respect to your frame in which they are separated by a distance L will differ in time by t in your frame. Because of that c squared in the denominator, the times involved tend to be very small. But at speeds close to that of light, the effect can take on great significance, as we shall see in the "twin paradox" example at the end of this chapter.

SPACE-TIME: THE FOURTH DIMENSION

Einstein's first few papers on relativity explored the subject so thoroughly that by the time others took notice of his work there was little left for them to add. One outstanding exception was the work of Hermann Minkowski, who had been one of Einstein's teachers in Zurich.

Minkowski was troubled by one feature of Einstein's approach: a single reality gives rise to multiple descriptions. Surely there must be another way to describe things, one that gets closer to the underlying reality. He found that he could do this by treating time *as if* it were a fourth dimension of space.

The words *as if* are crucial. Minkowski never meant to imply that space and time have lost their separate identities. He merely found that the postulate of relativity implies a *connection* between space and time that is analogous, but not identical, to the relation between different space dimensions.

In space alone, dimensions are connected via the *Pythagorean theorem*. If something is 3 miles east of here and 4 miles north, we can calculate the distance to it as the crow flies: $\sqrt{3^2 + 4^2} = \sqrt{9 + 16} = \sqrt{25} = 5$ miles. If the object is also at a different altitude than we are, just add in the square of that third dimension.

This calculation works only when we measure "north," "east," and "up" in the same units. So the first thing we must do to treat time as a fourth dimension is to measure it in units of distance. That is simple enough to do—we simply multiply by c to get the distance light would travel in that time. Astronomers sometimes use the reverse conversion, expressing the distance to a star in light-years.

Minkowski found that time can be appended to the Pythagorean theorem, but in a peculiar way that reminds us that we are not dealing simply with another dimension of space. Instead of *adding* its square to the squares of the space dimensions, we must *subtract* it! If we do this, we get a quantity that remains *the same in all reference frames*.

To illustrate this, let us return to the measurement of the length of the train in Chapter 9. We indicated that in the train frame, both telegraph poles were observed at the same time, but in the ground frame there was a time lag between the two measurements. As a result, the distance L between the two poles was greater than L_0, the length of the train at rest. By Minkowski's rule, in the ground frame we should subtract the square of the time interval from the square of this distance. In the train frame, the time interval is zero, so the space-time separation S is equal to L_0:

$$S^2 = L^2 - (ct)^2 = L_0^2$$

That way, both observers agree that the distance between the two observations, in four-dimensional space-time, is L_0. To the conductors on the train, the observations were simultaneous, and the distance is the length of the train. What the conductors call a pure length the signalman sees as a combination of a length and a time interval.

If the train is 1000 meters long, and moving at $0.6c$, the separation of the poles in the ground frame of reference is $\gamma \times 1000 = 1250$ meters. Using the formula from the last section, the time separation $t = 1250 \times 0.6/c = 2.5$ μs (microseconds). Converting this to a time by multiplying by c, it corresponds to 750 meters. Thus the space-time separation as calculated by the signalman is

$$S = \sqrt{1250^2 - 750^2} = \sqrt{1,000,000} = 1000 \text{ m}$$

the rest length of the train.

The simplest element of reality in this four-dimensional world is the *event*, something that happens at a particular *place* and *time*. The train conductors' observations of the two poles were two such events. The combined four-dimensional space-time separation of two events is the same in all reference frames, and is thus called an *invariant*. In the train frame, the two sightings were simultaneous, so the time separation is zero and the events are separated by the rest length of the train. In the ground frame the signalman must subtract the square of the time interval between sightings, the t in the preceding section, from the square of the distance between poles. Then he too gets the rest length of the train. The "contracted" length of the train is peculiar to his frame of reference, and is not an invariant, so it is not part of our four-dimensional picture.

Of course, we pay a price for this convenience. It is almost impossible to visualize a four-dimensional world. It can be dealt with only symbolically, through the medium of algebra. For this reason Einstein, who depended so heavily on visualization, initially rejected Minkowski's idea, but as he came to see its advantages, he embraced it wholeheartedly and made it the focus of his attempts to extend his theory to new phenomena. The fruits of this union will be explored in Chapter 12.

One factor that helps make relativity so baffling is the enormous disparity between our senses of space and of time. The eye can look upon objects a few inches away or gaze upon a grand vista covering tens or at most hundreds of miles. But we can scarcely imagine the time it takes light to cover such distances, which is less than a thousandth of a second. Thus there is no practical purpose to be served by reminding ourselves that we are not seeing remote objects now. But if we daily dealt with things moving at nearly the speed of light, we would ignore the time lag at great peril, and relativity might seem natural to us.

Our customary view of reality is like a motion picture—a series of still frames showing events in different places at successive instants in time. The problem is that two different movies are constructed in two different reference frames. Both contain all the same events, but those that are in the same still picture in one reference frame may be in different pictures in another. By adding the fourth dimension, we put every event in one big (but admittedly abstract) picture, with the space-time relations between things properly represented.

Treating time as a fourth dimension should be recognized for what it is— simply a bookkeeping device that better enables us to keep track of fast moving objects. The garage paradox reveals that neither what we see, nor the three-dimensional picture we construct from it by correcting for the time lags, is entirely satisfactory. In the four-dimensional picture, regardless of which frame we choose the door openings and closings are separated by the same distance, a bit less than 20 feet of space-time.

If we accept this convention, the speed of light becomes no more than a conversion factor between units of space and time, much as 2.54 is the conversion between centimeters and inches. This is now so widely accepted that it has become enshrined in our system of weights and measures. At present, we can measure time to an accuracy of about one part in 100 trillion. No measurement of distance can come close to this precision. Accordingly, a separate standard of length has been abandoned. The International Bureau of Weights and Measures simply took the best value for the speed of light, 299,792,458 m/s, and made *it* a standard. The meter is now defined as *the distance light travels in 1/299,792,458 of a second*. Michelson might be disappointed to learn that there is now no reason to ever again carry out his favorite experiment, measuring the speed of light. Its value has been fixed, once and for all, by international agreement.

ADDING VELOCITIES

Galilean relativity gives a simple rule for translating velocities from one reference frame to another. It is merely a matter of addition. If an object is moving at velocity u in a reference frame that is moving at velocity v with respect to you, you will see it moving at velocity

$$U = v + u$$

Obviously, this formula cannot apply in relativity. Among other things, it could allow U to become greater than c. The corresponding relativistic formula must take into account both the time and length translations. The result turns out to be

$$U = \frac{v + u}{1 + uv/c^2}$$

Two important consequences of this formula are worth noting:

1. If either u or v is small compared to c, the Galilean formula is nearly correct.
2. If $u = c$, then $U = c$, as can be shown by making this substitution and multiplying both numerator and denominator by c. This is of course what the postulate of relativity says—if a velocity is c in one reference frame, it must be c in every reference frame.

If both u and v are $\frac{1}{2}c$, the formula tells us U is not c, as in the Galilean case, but 0.8c. Similarly, in the examples we have studied of rockets moving at 0.6c with respect to one another, in the frame in which both are moving with equal but opposite velocity this velocity is not 0.3c but $\frac{1}{3}c$.*

---

*This can be shown by setting $v = \frac{1}{3}c$ and $u = -0.6c$ in the formula, which gives $U = -\frac{1}{3}c$.

THE TWIN PARADOX

One of the most startling predictions of relativity is illustrated by the following science fiction story.

A young astronaut takes a trip to a star twenty-five light-years away in a spaceship that can travel at 99.98 percent the speed of light, giving him a Lorentz factor of 50. The astronaut has a twin brother, who remains home on Earth. Fifty years pass, and the Earthbound twin, a graybeard bent with age, goes to the spaceport to welcome his adventurous brother. The astronaut bounds down the gangplank, because for him only one year has elapsed, and he is still young and vigorous!

From the point of view of the twin on Earth, this is because time itself slowed down on the ship. Clocks, and biological aging processes, are slowed to one-fiftieth their normal speed on this fast spaceship. To the astronaut himself, however, things seemed perfectly normal. From his point of view it was the Earth and the star that were moving, so the distance between them shrank to half a light-year, for a one-year round-trip at close to the speed of light. But both agree that the astronaut is now forty-nine years younger than his twin!

Early in the history of relativity, this story was offered as a refutation of the theory. Why isn't the twin on Earth the younger? After all, from the point of view of the astronaut, it is the Earthbound clock that ran slower! There appears to be a contradiction.

The answer is that one can make a distinction between the astronaut and his brother. The astronaut had to leave the Earth, accelerate to a stupendous speed, and turn around (another period of acceleration) at the star. Thus, he is not in uniform motion at constant speed, and the symmetry of relativity, by which he feels the Earthbound clock runs slow, does not apply. Analyzed in detail, the problem reveals that from the point of view of the astronaut, *most of the fifty years passed on Earth during the short time he was turning around at the star.*

That this is the resolution of the paradox can easily be seen if we imagine that, at the star, there is an unmanned artificial satellite, sent there from Earth in advance of the trip, with a clock set to "Earth time." Since the astronaut regards this clock and one on the Earth as moving clocks, they are indeed running very slow. But since the one near the star is far back along the line of apparent motion, it is also set ahead of the Earth clock. In the formula $t = Lv/c^2$, L is twenty-five years times c, and v is only slightly less than c, so the numerator is $25c^2$. In the astronaut's view the clock is nearly twenty-five years ahead of Earth time.

When the astronaut reaches the star and fires his rocket to stop and then turn around, he reverses the situation. The two "Earth time" clocks are now headed in the opposite direction, with the one on Earth trailing. Thus Earth time is now twenty-five years *ahead* of the clock at the star. If the remote clock on Earth switched from being almost twenty-five years behind to almost twenty-five years ahead, nearly fifty years must have passed on Earth in the brief time he spent turning around. Thus, the astronaut agrees with his twin brother: more time has passed on Earth than on the spaceship, and he is now some forty-nine years the younger of the two!

This fifty-year "leap" in Earth time came about because the astronaut *changed reference frames* in midvoyage. The theory of relativity as it stands has no way to explain this peculiar effect. In Chapter 12 we will come back to it again, for it represents one of the loose ends that drove Einstein to go beyond the theory we have studied so far.

Today, taking into account the twin paradox has become part and parcel of our normal technology of time-keeping. The cesium beam atomic clocks that are the basis for the world's time standards are accurate to a few nanoseconds a day. But radio time signals can be trusted to only about a tenth of a millisecond, because of uncertainty in estimating the distance traveled by radio waves, which in our atmosphere do not exactly follow straight lines. To keep other clocks in time with the world standards, portable atomic clocks have been flown on ordinary commercial airliners. With the help of a careful log of the plane's course, corrections are made for relativistic time changes, which can amount to tens of nanoseconds on a long flight. To test this procedure, in 1972 the U.S. Naval Observatory flew portable clocks on a round-the-world commercial flight, and compared them with clocks that had stayed in Washington, providing a direct experimental confirmation.

The literary possibilities of the twin paradox have been fully exploited by science fiction writers. One series of stories visualized an age when humanity has spread to habitable planets scattered throughout the galaxy, a civilization spanning thousands of light-years. In their powerful, speedy spaceships, a breed of astronauts maintains the skimpy "commerce" of this vast civilization, condemned to a strange existence in which they return to a familiar port only after centuries or millennia have elapsed, thus enjoying a peculiar sort of alienated immortality within a normal life span.

Summary

Relativity offers two remedies for conflicting pictures in different reference frames. One is to provide rules to *translate* from one frame to another, while the other constructs a new mode of representation on which all can agree. The translation rules involve the Lorentz factor *g* introduced in Chapter 8. Examples of the time and length translations show that these give pictures that differ only in unobservable features. The new mode of representation is to *formally* treat time somewhat as if it were a fourth dimension of space. This means two things: converting times to distances by multiplying by the speed of light, and using the Pythagorean theorem. The distinction between time and space remains because the time dimension is subtracted, rather than added, in the Pythagorean sum of squares. If this is done, the total "space-time distance" between two events is the same in all frames. The celebrated "twin paradox," in which an astronaut takes a long round-trip in space at close to the speed of light and winds up significantly younger than his stay-at-home twin brother, is analyzed. It turns out to be crucial that the astronaut *change* his reference frame in midvoyage.

CHAPTER 11

$E = mc^2$ and All That

What is matter?—never mind.
What is mind?—it doesn't matter.
—Anonymous

Up to now, we have been obliged to pay a heavy price for Einstein's simple postulate. If we embrace relativity, as experiments seem to say we must, we must learn to put up with a great deal of outrage to our common sense. We have no choice but to radically revamp our notions of space and time. Since these concepts are the very basis of the description of motion, it is reasonable to expect equally radical changes in Newtonian physics. Must we discard two centuries of progress, sweep away Newton's laws, energy and momentum conservation, and the like?

It turns out that the answer is no. This should hardly be surprising, for we must remember that Einstein had set out to *preserve* a feature of Newtonian physics that was under assault—the equivalence of inertial reference frames. Thus, most of the Newtonian edifice emerges from this part of Einstein's revolution intact.

Intact, yes, but by no means unchanged. And the major changes center on the concept of mass. Newton's first and third laws are obviously untouchable; unless we insist that mechanics concern itself with changes of motion that come about through mutual interactions of bodies, nothing recognizably Newtonian would survive. We will find that momentum conservation and the second law can be rescued simply by allowing mass to depend on velocity in a way that is now familiar, through the Lorentz factor. Moving on to energy conservation, we will find that this result leads to the duality between mass and energy expressed in the most widely publicized formula of twentieth-century physics, $E = mc^2$.

But the arguments in this chapter should be far less of a strain on your credulity than those in the one preceding, because their consequences will not seem quite so outrageous.

THE MASS INCREASE

The first task is to establish the dependence of mass on velocity, through a *gedanken* experiment based on a grazing collision between two identical space-ships. This kind of collision is familiar to all billiard players. The moving object is hardly deflected and loses little speed, while the struck one comes out at a slow speed nearly at a right angle to the original motion. How this looks to the spaceship crews is shown in Figure 11-1.

   The trick to understanding this situation is to consider only the part of the motion that is perpendicular to the original line of motion. In this direction we need not concern ourselves with disagreements in measurements of length, but need only consider the moving-clock effect. And Galileo gave us the right to study components of motion on perpendicular directions separately.

   We now ask what we must accept in order that momentum will continue to be conserved in the perpendicular direction. In order to test this law, let each crew start a timer at the moment of collision, and time how long it takes to drift a given distance x in the perpendicular direction. From the symmetry of the situation, it is clear that both crews will get the same value for the time t, and thus report the same perpendicular component of velocity. Since both

FIGURE 11-1.

Collision as seen by observer at
rest with respect to body *a* before
the collision   (observer A)

Same collision from point of view of observer
at rest with respect to *b* (observer B)

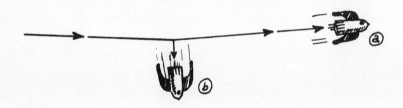

ships have the same mass, both crews will get the same value for the momentum, mx/t.

It would seem as if we are home free, but we must remember that neither crew accepts the time measured on the other's clock as valid for its own reference frame. Each immediately multiplies t by γ. Each thus adds a γ to the denominator of the momentum of the other ship, destroying the balance in momentum. To restore it, they also need a γ in the numerator. Since both crews agree on the distance x, there is no place to put this factor except on the mass.

The conclusion is that momentum conservation holds only if a fast-moving body behaves as if it has more mass than a slow-moving one. The same argument works along the direction of motion, but the calculation is more difficult. So when we speak of mass, we must introduce the symbol m_0 to designate the mass of an object at rest. The relativistic mass* is then defined as

$$m = \gamma m_0$$

Like the rest length L_0, the rest mass m_0 is an invariant. The mass m is correct for only one reference frame.

This modification does not change the essence of the concept of mass, which is "the tendency to resist a change in motion." It should not puzzle us unduly to find that fast-moving objects are more resistant to a change in motion than slow ones. Moreover, this effect ties up a loose end that has been dangling since Chapter 9: how do we enforce c as an absolute speed limit?

NEWTON'S LAWS AND THE RELATIVISTIC SPEED LIMIT

In Chapter 3 we saw that Newton's second law defines *force* as the rate of momentum transfer. Because changes in momentum normally involve changes in velocity with no change in mass, we could write the formula as $F = ma$. But in relativity, both mass and velocity change. Newton's definition still holds, but the simple formula is reduced to the status of an approximation that holds at low speeds, where γ does not change rapidly.

Let an object be accelerated from rest by a constant force. At first, the mass changes very little, so all the increase in momentum comes about through changing the velocity. But as it begins to approach the speed of light, the object becomes more and more massive and more and more resistant to a change in velocity. Momentum is still increasing at the same rate, but most of it goes into a change in mass, with very little change in velocity.

As we approach the speed of light, the Lorentz factor approaches infinity. The mass becomes an impassable barrier to further acceleration. In modern particle accelerators, subatomic particles such as electrons or protons are acted on by powerful electromagnetic forces. As of 1996, the record speed for a particle in one of these machines was $0.999999999987c$, achieved at the LEP (Large

---

*Some relativity texts reserve the term "mass" for rest mass alone and put the γ in the definition of momentum. This is simply a semantic choice.

Electron-Positron) accelerator near Geneva, Switzerland, a 17-mile ring of magnets and vacuum pipes that cost more than a billion dollars to build. If simply pushing hard enough or long enough could do the trick, the speed limit would have been violated long ago.

One might reasonably ask just what is that particle accelerator *doing?* Surely it is not worth all that money and effort to get a tiny increment of speed. The answer, of course, is that the electric field is still transferring momentum to the particle—and it is also doing work. Work transfers *energy.* And the visible signs of that energy are the increase in the particle's *mass.* The electrons whirling around LEP are 200,000 times heavier than an electron at rest.

This points the way to a deeper, more general result, of which the increase of mass with velocity is only one special case: *all forms of energy have mass.*

ENERGY AND MASS

It can be shown that the work done on an electron in LEP is just the change in its mass multiplied by c squared. But energy conservation transcends motion itself. Let us therefore establish the relation between mass and energy by a *gedanken* experiment that emphasizes the universal character of energy conservation.

As we saw in Chapter 6, electromagnetic forces can conserve momentum and energy only if we ascribe these qualities to the field (and hence to light) itself. If an object emits light in one direction, in order to conserve momentum it must itself "recoil" in the opposite direction. Maxwell's theory leads to a simple relation between momentum and energy for light, $E = pc$. From the definition of momentum, we get $p = mv$. Since the velocity is that of light, we can write $p = mc$ and substitute this value in the Maxwell relation, obtaining

$E = mc^2$

Since it is light we are dealing with, what is the meaning of this mass? Is it more than a mathematical fiction? Consider the situation depicted in Figure 11-2. In a closed box floating in space, a flash of light is produced by a battery-operated flash lamp at the left end. If this process conserves momentum, the box must recoil ever so slightly to the left. Then let the light flash be absorbed at the right end. The momentum of the light is transferred back to the box, bringing it to a halt.

Without any external force acting on it, an initially stationary box was displaced to the left. This is exactly what would happen if a mass inside the box moved from left to right!

One might argue that this merely proves that light can be used to transfer mass, not that there is any general relation between mass and energy. But consider the state of the box after it stops. The light no longer exists. None of the actual material of the box has been removed from the left and transferred to the right, yet mass has been transferred between the two ends! What other changes have taken place? What evidence remains of the transfer of the light?

Closed box at rest...

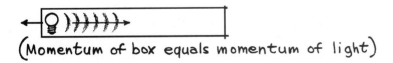

begins to move to left when light flashed
at left end

(Momentum of box equals momentum of light)

Stops when light absorbed at right end...

Thus light has transferred mass from one
end of the box to the other

FIGURE 11-2.

The answer is that the absorbed light heated the right end of the box. Conversely, at the left end, energy was removed from the battery. Some chemical energy at the left end of the box was transformed into electricity, and then into light, and finally into heat at the right end. At the same time, mass was transferred from the left to the right! Thus, all these forms of energy must obey a mass-energy relationship. A discharged battery must be lighter than a charged one, a hot object heavier than the same object when it is cold. But since heat is nothing but the energy of motion of molecules, this too must obey a mass-energy relationship. And through the law of the conservation of energy, the mass-energy equivalence can be extended to any form of energy whatsoever.

It must be emphasized that $E = mc^2$ is the one and only formula for energy in relativity. What then has become of our old definition of kinetic energy, $\frac{1}{2}mv^2$, which still ought to work at low speeds? The answer is that the kinetic energy represents merely the tiny increase in mass of a slow-moving body. Remember the approximate formula for γ, valid at low speeds, in Chapter 8. Multiplying by m_0c^2 we get

$$\left(1 + \frac{1v^2}{2c^2}\right) \times m_0c^2 = m_0c^2 + \frac{1}{2}mv^2$$

The m_0c^2 term is called the *rest energy* of the object, the energy it has simply by existing. It reflects the fact that it takes energy to create matter. The second term is the familiar definition of kinetic energy, the additional energy due to motion.

It is reasonable to ask how light can move at c when it has mass. The answer is that *all* of light's mass is "kinetic"—it has no rest mass at all.

THE MEANING OF $E = mc^2$

Ever since Hiroshima, this formula has been associated in the public mind with nuclear energy. At the risk of repetition, it must be stated emphatically that it applies equally well to all forms of energy; it is a quite universal and unique formula, as valid for a bonfire as for a nuclear weapon. The only distinction of nuclear energy is that it is the one energy source powerful enough for the changes in mass to be really substantial.

Heating water from its freezing point to its boiling point increases its mass by only about one part in 100 billion, or 10^{11}. In an ordinary chemical reaction, such as a fire, the combustion products are lighter than the fuel and oxygen used in the fire by about one part in 10^9. Such small changes are beyond the range of measurement. But in the more violent nuclear reactions, mass changes of about a part in a thousand can take place. From a table of nuclear masses, a physicist can use Einstein's formula to predict the energy release of a previously unstudied reaction.

The crucial factor for the development of nuclear energy and nuclear weapons was the discovery of the nuclear fission chain reaction, in which each disintegrating nucleus triggers several neighbors. The formula $E = mc^2$ gave no hint of the existence of any such reaction, nor is it essential to the understanding of the process. Einstein himself played no role in the discovery of fission or in the development of the atomic bomb, aside from the initial one of affixing his signature to a letter to President Franklin D. Roosevelt, drafted by other scientists, warning of the possibility of an atomic bomb and of evidence for a German effort to build one. Had relativity not yet been discovered, it would probably have hampered the efforts of the Manhattan Project very little.

The formula is sometimes mistakenly referred to as a formula for the conversion of energy into mass. It is more than that; it is a statement that, for all practical purposes, the two are identical. All forms of energy have mass, and what we used to call mass, which we now call rest mass, is just one more form of energy. This is a perfectly natural extension of nineteenth-century ideas of energy conservation.

For example, if a bonfire occurs in a sealed box, insulated so that the heat cannot escape, no change in its weight can occur. Despite the transformation of chemical energy into heat, which represents kinetic energy of the molecules, no change in mass has taken place. If we allow the heat to escape, however, the box will become slightly lighter.

As a last example to illustrate the generality of the mass-energy equivalence, let us apply it to potential energy. Whenever a force binds two objects together, their combined mass is less than the sum of their separate masses. The negative potential energy appears as what is called a *defect* in mass. This is the real source of nuclear energy: the potential energy of the powerful forces that bind the component particles of the nucleus together. A rearrangement of a large, loosely bound nucleus into smaller, more tightly bound ones increases the strength of the binding, lowering the mass of the nuclei. Since this mass is field energy, it is distributed in space. Thus, as the concept of field matures in physics, it acquires more and more of the qualities we associate with matter.

A typical nucleus weighs about 0.8 percent less than the combined mass of its component particles. The atomic bomb derives its awesome power from the fact that the heaviest nuclei weigh about 0.1 percent more per particle than nuclei near the middle of the table of elements, and from a process (fission) that splits certain heavy nuclei into two smaller parts, releasing that one part in a thousand of their mass in the form of heat.

Outside the nucleus, the mass defects are much smaller. A hydrogen atom, for example, weighs about one part in 600,000 less than the proton and electron of which it is made. And the solar system weighs about one part in a trillion less than the combined mass of the Sun and the planets and other bodies that circle it.

The energy-mass equivalence once again illustrates the role of the velocity of light as a conversion factor between quantities that were originally regarded as distinct. Had the equivalence of mass and energy been understood from the outset, there might not have been separate units for the two. Today, physicists working with subatomic particles, where grams are horrendously large and awkward units, use energy units for mass as a matter of course. Viewed in this context, c squared has no more profound significance than the conversion factor 0.621 from kilometers to miles.

The square of the speed of light is, however, a rather *large* conversion factor. Energy doesn't weigh much. To get just 1 kilogram of energy, you need $(3 \times 10^8)^2 = 9 \times 10^{16} = 90,000,000,000,000,000$ joules! The bombs that destroyed Hiroshima and Nagasaki lost about 1 gram of mass each to produce their devestating results.

One kilogram of energy is in fact just about what our larger electrical generating plants produce in the way of heat in a year (only about a third of this heat is converted to electricity). If the plant is nuclear, that kilogram comes from the fission of about 1 ton of fissionable material, which is part of a fuel assembly weighing perhaps 50 tons. If instead it is a coal-fired plant, it must consume nearly a million tons of coal—more than 100-mile-long trains of hopper cars each year!

The dependence of mass on velocity was one of the first predictions of relativity to be experimentally confirmed. Electrons are so light that it is rather easy to accelerate them to considerable velocities. Those in a typical TV color picture tube travel at nearly one-fourth the velocity of light and have a mass nearly 3 percent greater than when standing still. And even higher velocities are easy to obtain. A measurement of the mass increase of electrons was performed as early as 1906. The energy-mass relation in nuclear reactions was confirmed to high precision in 1932 in the first artificial disintegration of a nucleus.

Today, experimental confirmations of all aspects of relativity are commonplace. Physicists studying subatomic particles work daily with objects traveling close to the speed of light. In particle collisions, these particles bear out all the details of Einstein's predictions. For example, some particles are highly unstable and break up spontaneously in the time it takes for a light signal to travel several centimeters. Yet, at close to the velocity of light, they can be transported many meters with no difficulty, because of the slowdown in their internal "clocks."

From the point of view of the particle, it is not the slowing of time but the contraction of the laboratory that is responsible for the effect. Yet the end result is the same: it reaches the detector. And when a particle that would be very light when standing still is brought up to a high speed, it acts like a heavy particle in collisions. Finally, the extra mass obtained by accelerating a particle close to the speed of light can be used to produce new particles not previously present.

The final triumph of relativity was to complete the work of Faraday and Maxwell in uniting electricity and magnetism. Einstein was able to show that a magnetic field appears when a purely electric field is seen by a moving observer, and an electric field appears when a purely magnetic one is seen from a moving vantage point.

A RELUCTANT REVOLUTIONARY

Einstein had a deep respect for Isaac Newton, and was by temperament a most reluctant revolutionary. The 1905 papers on relativity saved Newtonian physics by reformulating it in a manner consistent with Maxwell's theory of light. But Einstein went on to a deeper and more revolutionary insight. Thereafter, he called the theory we have studied so far *special relativity*. In 1915, after a struggle of eight years, he presented to the world a theory he called *general relativity*. This extended the relativity principle to *accelerated* reference frames, and brought gravity within its scope.

Einstein's goal was a complete reformulation of physics in which the concept of *force* disappears altogether. Instead, gravity is described as a distortion of the very fabric of space itself, until a straight line is no longer the shortest path between two points.

It is general relativity that gave birth to the oft-cited (but decidedly false) legend that only twelve people in the entire world could fathom what Einstein was talking about. Hundreds of physicists and mathematicians understood the theory in Einstein's day, and thousands do today. Though it is based on formidably difficult mathematics, the basic principles of general relativity are neither difficult nor obscure. They are the topic of Chapter 12.

Summary

Despite relativistic reinterpretations of space and time, momentum conservation survives if we simply allow mass to vary with velocity according to the Lorentz factor. This automatically enforces the relativistic speed limit, for mass would be infinite at the speed of light. Through analysis of light in a closed box the energy-mass relation $E = mc^2$ is obtained. Though popularly associated with nuclear energy, this formula applies to all forms of energy, and simply means that "energy has mass." It is a natural extension of energy conservation.

CHAPTER 12

Did God Have Any Choice?

Within every creature incarnate sleeps the Infinite Intelligence unevolved,
hidden, unfelt, unknown—yet destined from all eternities to waken at last,
to rend away the ghostly web of sensuous mind, to break forever its
chrysalis of flesh, and to pass to the extreme conquest of Space and Time.
      —LAFCADIO HEARN, The Diamond Cutter

At some time in the fall of 1907 Albert Einstein, by then technical assistant second class at the Swiss Patent Office, had what in later years he would call "the happiest thought of my life." As he explained it,

> I was sitting in a chair in the Patent Office in Bern when all of a sudden a thought occurred to me. "If a person falls freely he will not feel his own weight." I was startled. This simple thought made a deep impression on me. It impelled me toward a theory of gravitation.

Nothing immediate, however, came of this happy thought, for he would soon find that between him and his goal lay eight years of false leads and fresh starts, as well as mastery of some of the most daunting mathematics ever invented.

All Einstein had done, of course, was to "rediscover" Galileo's law of falling bodies: all of them fall with the same acceleration. If one falls freely— or orbits the Earth, which is a form of free fall—in a closed capsule, everything in the capsule falls with the same acceleration, leaving no relative motion to indicate the presence of gravity. Today, with live telecasts from manned space-craft, we are all familiar with this effect, but in Einstein's day it took a stroke of the imagination.

Newton's way of dealing with this peculiar property of gravity had been to make the force proportional to mass. But Einstein had redefined mass some-what, into a quantity that was different in different reference frames, and was not yet sure what this might do to Newton's law of gravity. Furthermore, the law of gravity needed to be reformulated from Newton's instantaneous action at a distance to a field whose propagation was limited to the speed of light. Finally, gravitational field energy itself now had mass. It was not yet clear how to deal with these matters, but at least he had a test that any new theory must pass: it must preserve Galileo's law. And as he had done with the postulate of relativity, Einstein once again chose this fundamental rule not as the *goal* of his theory but at its *logical starting point!*

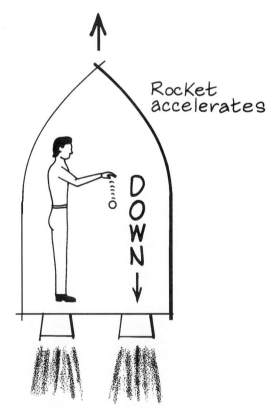

FIGURE 12-1.

Einstein's happiest thought leads naturally to its converse: in a closed capsule moving with constant acceleration, it will seem as if a gravitational field is present. Consider the rocket in Figure 12-1, which is out in interstellar space, accelerating with its rockets firing. Any loose object in it will "fall" toward the tail with uniform acceleration. Of course, in Newtonian terms, no force is acting on the object; it is really the rocket that is accelerating. Thus all objects will have the same acceleration—that of the rocket. Galileo's law is automatic in an accelerated reference frame.

Thus Einstein set out to construct a single theory that would apply equally well to both accelerated reference frames and gravitational fields.

THE PRINCIPLE OF EQUIVALENCE

To develop his theory of gravity, which he called *general* relativity, Einstein added one more postulate to the theory of relativity:

No experiment performed in one place can distinguish a gravitational field from an accelerated reference frame.

The words "in one place" are important. Einstein did not eliminate *all* distinctions between gravity and accelerated reference frames. If we compare the fall of objects in different places on Earth, we will find that they all head toward the Earth's center. No accelerated reference frame can duplicate that pattern, though a rotating frame can have acceleration directed *away* from a central axis. What Einstein is saying is simply that all objects respond to a gravitational field in the same way they respond to being in an accelerated frame.

As with the special theory, this seemingly innocent hypothesis, which is known as the *Principle of Equivalence,* turns out to have unexpected consequences. Before we explore them, let us spell out our destination and the route we shall take to it.

Our conclusion will be that there is absolutely no need for a "force" of gravity. The acceleration of falling bodies or planets is simply a case of inertial motion, of objects coasting along on the shortest paths available to them. But these paths are not straight lines, because *space-time itself is not flat!*

The problem with these words is that we can scarcely imagine what they mean. It is hard enough for us to accept the fact that the surface of our Earth is curved. Looking at a flat map, it is hard to visualize that the curved paths we call great circle routes or, to use their mathematical name, *geodesics,* are really the shortest routes between points on the Earth's surface.

But at least we can be reassured that the Earth sits in a perfectly flat space, and if we could tunnel through the Earth, that would be the shortest route. But space itself curved? What do the words mean?

We shall try to remove some of their mystery through the tried-and-true device of *gedanken* experiments. These are designed to show in turn:

1. In an accelerated reference frame, space-time is curved in a trivial way.
2. Light moves in curved paths in accelerated frames and thus also in gravitational fields.
3. If we accept the above as meaning that space-time itself is curved, we can explain everything we know about gravity without ever having to mention a force.
4. Gravity affects time, turning the fifty-year time leap in the twin paradox example into a real effect.
5. People who live in round worlds but insist they are flat are bound to invent forces like gravity.
6. There are ample experimental tests of all the above, and some important effects that can be explained in no other way.

The payoff for climbing this path will be a physics that contains some fascinating curiosities like black holes, and gives insight into the origin of the universe itself.

The rules of the road are simpler than for the arguments we used to develop special relativity. We must simply accept that once we have demonstrated that something happens in an accelerated reference frame, it must also happen in a gravitational field.

THE WARP AND THE WOOF

Galileo could have drawn Figure 12-2 if he had known about graphs, which were invented by Descartes. But it is more than a graph; in our new four-dimensional language, it is a map of a two-dimensional slice of space-time, showing the path of an object moving freely in an accelerated reference frame or a gravitational field. It is obviously a curved line.

Newton would immediately ask whether there is in fact a true gravitational field present. If so, the path is really curved. If not, we are seeing what is really a straight-line path distorted into a curved one by viewing it from an accelerated reference frame. For Einstein, this distinction is meaningless. In either case, the object is obeying the Law of Inertia, as it applies in the local reference frame: inertial paths are curved here.

Now let us extend the principle to light. Suppose we are inside the rocket in Figure 12-1 and shine a laser beam across it from wall to wall. Since the opposite wall will speed up while the light is in transit, the beam will hit that wall to the rear of the rocket from the point it was aimed at, and the path of light across the rocket will in fact be a parabola. The Principle of Equivalence forces us to concede that the same thing must happen in a gravitational field.

The curvature is so small that we would have a hard time proving this. Suppose that on a very clear day, we shine a beam from one mountain peak to another, 20 miles distant. It would get there in a ten-thousandth of a second. In that short time, a falling body drops by only a distance not much larger than the diameter of an atom! We could never distinguish the path of a beam that curved that little from a straight line.

There is nothing in this example that would have caused Newton to lose a night's sleep. In his second most celebrated work, *Opticks*, he raised the possibility that gravity could affect the path of a light beam. The real argument is about how we describe the *cause* of what we see. So let us move on to an ordinary falling body and see how Einstein accounts for its motion.

FIGURE 12-2.

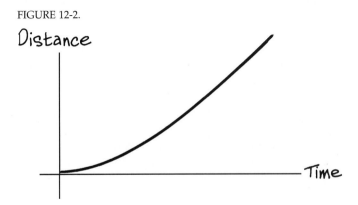

You may now object that such a slight curvature could hardly account for the motion of falling bodies. But we must remember that the curvature is not of space alone but of space-time. The fact that we do not notice the curvature is yet another consequence of the disproportion between our sense of space and our sense of time.

If you drop a coin from waist height, it reaches the floor in about a half second. While we regard this as a small interval of time, its space equivalent for our four-dimensional picture is immense. In that brief time a light beam covers about 100,000 miles. The equivalent of Figure 12-2 for that coin is a graph with vertical dimensions of a few feet but horizontal dimensions of 100,000 miles. We have sampled a huge two-dimensional slice of our four-dimensional space-time, and its tiny curvature is quite sufficient to explain the motion of the coin.

Einstein says that when the coin is released, it simply follows its natural curved path in space-time. This brings it, after its long journey, in contact with the space-time path of a point on the floor. This other space-time path is straight, because gravity is not the only game in town. Whatever gives solid matter its hardness must also be able to influence the geometry of space-time. It can overcome gravity and straighten out the floor's space-time path.

There is no quarrel between Newton and Einstein over the description of these space-time tracks. What they disagree about is their significance. Newton says the tracks are curved by the action of a force. Einstein insists that no force is necessary: space-time itself is curved.

We have now completed the first three (and the easiest) steps of our argument, and still we have no way to choose between the Newtonian and relativistic explanations of gravity. The last three will remedy this situation.

DOES ANYBODY HAVE THE RIGHT TIME?

We are now ready to resolve the quandary of the astronaut in the twin paradox. Figure 12-3 shows an accelerating rocket ship, equipped with a clock in its nose that sends time signals ten times each second, to be received further down in the ship.

But the rocket is accelerating, so that in the time it takes for the signals to reach the receiver, it has gained additional speed and is headed toward them. Suppose, for example, that it gains one-tenth the speed of light. Then adding this movement to that of the signals themselves will allow the receiver to pick up one additional signal each second. We then have a clock sending ten signals each second, but a receiver picking up eleven!

How is this possible? Surely we are not manufacturing additional signals along the way! We have only one way out, and that is to admit that *a second at the bottom of the rocket must be different from one at the top.* Since the receiver is picking up 10 percent more signals than the clock sends, its second must be 10 percent longer. This means that a clock at the bottom of the rocket actually runs 10 percent slower than one at the top. The Principle of Equivalence then insists that the same thing must happen in a gravitational field.

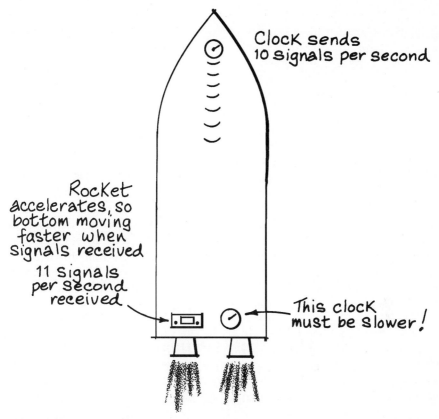

Clock sends
10 signals per second

Rocket
accelerates, so
bottom moving
faster when
signals received

11 signals
per second
received

This clock
must be slower!

FIGURE 12-3.

To see how this helps the astronaut get his story straight, let us find a formula for the clock slowdown, which is simply the ratio of the speed gained while the signal is in transit to the speed of light. If the acceleration is a and the height of the rocket h, over a time interval t the upper clock will get ahead of the lower one by

$$\delta t = \frac{ah}{c^2} t$$

In a gravitational field, a is simply the acceleration due to gravity, roughly 10 meters per second squared on Earth. That c squared in the denominator is a pretty big number, so the result is usually tiny. In a balloon at a height of 10,000 meters, a clock would gain only 4 nanoseconds an hour compared to a clock on the ground.*

---

* The formula is valid only for altitudes that are small compared to the radius of the Earth. Otherwise the inverse-square variation is important, and you should use $\delta t = (GM/rc^2)t$.

Small as this is, atomic clocks are good enough to measure it. It is also roughly comparable to the loss of time due to the speed of a commercial jet plane. Thus both effects must be taken into account when portable atomic clocks are flown around the world.

For the astronaut in the twin paradox, however, this effect would not be tiny at all. In order to reverse a speed close to that of light in a few days, his acceleration would have to be billions of times that of Earth's gravity, and the Earth quadrillions of meters away in the "up" direction. Thus during his turn-around, Earth's time would pass many times faster than his. Since firing his rockets in this remote location could not have had any effect on Earth time, he must conclude that time, for him, was essentially "frozen" during his turn-around, and fifty years passed on Earth in what seemed to him a very short period.

His stay-at-home twin can now only reply, "I think you're ridiculous to use such a silly reference frame, but I have to admit that your version of the story, bizarre as it may be, is at least self-consistent." Note that this is not a case of two observers each believing the other's clock is slow. Everyone agrees that in an accelerated reference frame the speed of a clock depends on its position and clocks higher "up" run faster.

THE FLAT-EARTHERS

Don't be disappointed if you still can't fathom curved space-time. After all, there are still people around who believe the Earth is flat!

Let two of these benighted individuals begin a journey due north, one starting 100 miles east of the other, traveling at the same speed. We who believe in a round earth know what will happen to them. They will draw inex-orably closer to each other, eventually meeting at the North Pole.

If they keep track of their separation along the way, they will find that ini-tially the approach is very gradual, but it speeds up toward the end; accelerated motion! If they persist in using plane geometry, they will be hard-pressed to explain what went wrong. One way would be to say they were drawn toward one another by a force. Just like gravity, their "accelerations" would be inde-pendent of their masses, so it would have to be a force proportional to mass.

This is just a simplified example, on the two-dimensional curved surface of our planet, of the consequences of failing to recognize the curvature of four-dimensional space-time. Since this example did not include a time dimension, we substituted the steady progress of our travelers for the familiar steady march of time. Otherwise, the analogy is exact.

But to arrive at this point, Einstein needed help, and it came from a famil-iar quarter. Curved geometry is one of the most difficult problems in math-ematics, and was far beyond Einstein's training. In 1912, when he returned from Prague to Zurich to take a professorship at his alma mater, he approached his old friend Grossmann, who by then was the professor of math-ematics. "Grossmann," he pleaded, "you must help me, or else I'll go crazy!"

Fortunately, sanity was near at hand. Fifty years earlier the mathematician Bernhard Riemann had constructed a geometry tailor-made to Einstein's needs, and had even suggested that it might be worthwhile to check the geometry of the space we live in on a large scale, to see if it is really Euclidean after all! Though Grossmann was no expert in this area, he at least knew Riemann's geometry existed. The two joined forces for just a few months, until Einstein's departure for Berlin once again broke up the collaboration. But that was enough to set him on the proper path.

The help proved crucial, for this time Einstein was not laboring in obscurity. Preliminary versions of his theories had appeared in print, only to be withdrawn when they failed to work. He was in a race with others, some of whom, such as the mathematicians David Hilbert and Emmy Noether, were far more adept than he at this sort of mathematics. Nonetheless, guided by superior physical insight and aided by his phenomenal powers of concentration, he beat them all to it.

NEWTON'S LAST STAND

We have already seen that the curvature of space near the Earth is far too small to have a measurable effect on a light ray. But the Sun's gravity is much stronger, at least up close. Figure 12-4 shows what happens to the image of a star when its light must pass near the Sun to reach us.

The curvature of space near the Sun is sufficient to bend light through an angle of 1.7 seconds. Small as this angle is, in a photograph taken by a telescope 20 feet long the star's image will be about 0.05 millimeter from its normal position, which is within the limit of measurement on good glass-backed photographic plates.

Newton, of course, would have expected a similar effect. Indeed in 1801, the astronomer Johann von Soldner used Newton's Law of Gravity to calculate the path of an object traveling at the speed of light that just grazes the surface

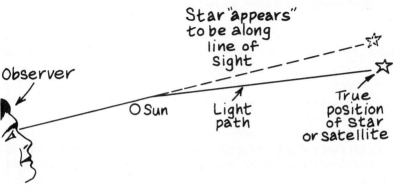

FIGURE 12-4.

of the Sun, getting a deflection of 0.84 second. Without knowing of this result, Einstein had duplicated it in 1911, from the bending of a light wave due to the gravitational clock effect in *flat* Euclidean space, for he had not yet discovered that space-time is curved. In curved geometry, the effect is twice as great.

Of course, to photograph stars near the Sun you must wait for a total eclipse and take your telescopes to it. On the strength of Einstein's reputation, a German expedition went to the Crimea to photograph a total eclipse in August 1914. They never got to take their pictures: by the time of the eclipse, Germany and Russia were at war and the astronomers were in temporary detention as enemy aliens.

A year later Einstein had his theory, and with it the new prediction. Had the German expedition succeeded, Einstein would have been guided to the right theory by the unanticipated failure of his earlier prediction, and his triumph would have seemed a lesser feat. Yet again, Einstein's luck had held out.

As World War I drew to a close, the British astronomer Arthur Eddington learned of Einstein's prediction through contacts in neutral Holland. Here was a chance for a historic head-to-head confrontation of two theories. He noted that in the spring of 1919 there would be an eclipse in the south Atlantic. Furthermore, it would take place in a part of the sky where the stellar backdrop was the Hyades, a dense cluster of stars bright enough to see during an eclipse, greatly improving the accuracy of the test.

Expeditions were sent both to Africa and South America—Eddington wanted to take no chances that bad weather would rob him of his prize. When the photographs were compared with others taken by the same telescopes earlier in the year, with the Sun out of the picture, Einstein's prediction was borne out.

Today, in the era of radio telescopes, we are far more fortunate, for our Sun is a comparatively dim radio star and we need no longer wait for an eclipse. The measurement has been repeated several times since the 1970s, and every new refinement confirms Einstein more precisely.

As soon as he was sure of the result, Eddington relayed the news to Berlin, again via Holland. It was received with Einstein's usual equanimity. When his secretary asked how he would have felt if it had turned out wrong, he replied, "Then I would have felt sorry for the dear Lord, for the theory is correct!"

An even more serious blow to Newtonian gravity came from a little spacecraft named *Mariner 9*, the first to orbit Mars, in 1972–1973. It carried a radar transponder that beamed back radar signals sent from Earth. When the line of sight to Mars passed near the Sun, another dimension was added to the situation in Figure 12-4, measuring the transit time of the radar signal, which is about twenty minutes each way.

In flat Euclidean space, it is easy to predict the result. Trace the curved path of the signal on a flat plane and see how much longer it is than a straight line. The answer is about 30 feet, so the signal would arrive only 30 nanoseconds later than would be expected from the known positions of Earth, Mars, and the spacecraft.

But in Einstein's curved space-time, this estimate is as useless as trying to get the exact distance from New York to London off a flat map. Near the Sun, time is slowed down, so a light signal is further delayed. The formulas of curved geometry give a delay of 200 microseconds, over 6000 times longer than the flat-space figure. Irwin Shapiro of MIT headed the team that confirmed the longer delay.

There are other theories of gravity, and Einstein's may not win the final beauty prize. But all the remaining contestants are curved-space theories. Flat space-time is dead.

GOODBYE PHYSICS, HELLO GEOMETRY

With the acceptance of curved space-time, the science of mechanics is reduced to utter triviality. Physics disappears, leaving nothing but a geometry. No laws of motion are needed any more, just a single statement:

All objects follow the geodesics of curved space-time.

Or, as the American theoretical physicist John Archibald Wheeler put it, "matter tells space how to curve, and space tells matter how to move."

But what a geometry! That part is anything but trivial. To get just a hint of how complex it is, consider the following. To specify the curvature of a one-dimensional line at some particular point takes but one number, the radius of curvature. Since a two-dimensional sheet can be bent in two directions, with any desired angle between them, to specify its curvature takes three numbers. Though we cannot visualize what curved three- and four-dimensional spaces are like, we do know how many numbers we need: six and ten, respectively.

Ten numbers! Newton's theory of gravity got by with just one, the force. Clearly, Einstein's is more complex. A terribly abstruse mathematical language, *tensor differential geometry*, is needed even to describe it.

It is so complex that it is useless in most practical situations. Its role in physics is like that of a sacred text, locked away from the multitude but consulted by high priests on sacramental occasions. It is used to verify formulas that apply to fairly simple situations or approximate ones that give the effects of curved space-time as small corrections to Newton's law of gravity.

Do not expect your intuition to be a reliable guide in a curved four-dimensional world. You would be bucking hundreds of millions of years of evolution of the human central nervous system. Our eyes are our most precious sense organs, and our brains are wired to process the information they deliver by the rules of Euclidean geometry. The only reason we can grasp the fact that the two-dimensional surface of the Earth is curved is because there is a third dimension for it to curve through. But it would be a mistake to assume that if four-dimensional space-time is curved, there must be a "fifth dimension" for it to curve through. Those ten numbers are enough to tell the whole story. We need no outside reference. It's all the space we've got, and it is fundamentally, undeniably curved.

THE TAO OF SPACE-TIME

The last link in our new geometric worldview is to plug in the mass-energy equivalence and see what wonders ensue:

1. Curved space-time means that a field is present.
2. Fields store potential energy.
3. Energy has mass.
4. Mass is the source of gravitational fields; so back to 1!

What a long way Faraday's little lines of force have carried us! They started as a way to avoid the problem of action at a distance. Now they generate their own matter. And at the same time, they are the very fabric of space-time.

We began with a physics that needed four kinds of reality; space, time, matter, and a cause of motion, first force and later energy. Special relativity forged links between space and time and between matter and energy. Now the unification is complete:

Matter is energy is space-time.

There is harmony enough in these nine syllables to please a Zen master. But while we admire the austere simplicity of this worldview, let us recognize it for what it is not as well as what it is. It is not a completed task but a commitment to a project, a frame in which to hang our picture of the universe. And that picture is by no means complete, for gravity is not the only field there is. Einstein spent most of the last twenty-five years of his life trying to build electromagnetism into his geometry, and in the end he failed.

The structure of matter is ruled by electromagnetism, and also by other fields that operate on the subatomic scale. We cannot understand the universe until we understand the atom.

Modern physics began with the study of gravity, and its first triumph was to resolve the dispute as to whether the Earth moved. Now, in the light of Einstein's theory, we can declare the whole argument moot! For Einstein, there are now no priviledged reference frames. At least in principle, it is possible to account for the apparent motions of the Sun and planets, not to mention the rest of the universe, in a frame of reference in which the Earth is at rest. In practice, however, the geometry of such a space-time would be monstrously distorted. Einstein would say "stick with your Sun-centered system—that way the geometry is closer to Euclidean." What had been a matter of principle has become a matter of convenience.

A COSMIC VACUUM CLEANER

There are two neat cases in which the geometry of space-time can be solved exactly. One of these is the universe as a whole, and the other is a large spherical mass, such as our Sun or the Earth. The latter case leads to one of general relativity's most bizarre predictions, which John Wheeler in 1963 dubbed the *black hole.*

The idea of a black hole is much older than its name. Laplace pointed out as early as 1796 that if a star contains enough mass in a small enough package, the velocity of escape from its surface is greater than that of light. No light can then get out, though light and matter can fall in.

In 1916, only a few months after Einstein completed the general theory, Karl Schwarzschild showed that the same effect should still be present. Of course, in general relativity the light does not just fall back. It simply travels on curved paths smaller than the size of the star. Since in relativity the speed of light is the limit for all speeds, nothing that is inside the black hole can escape. The star is, for all intents and purposes, plucked out of space-time.

The density of matter required is phenomenal. Our Sun would have to be only a few miles in diameter to become a black hole. The pressure generated by the nuclear "flame" in its heart prevents it from collapsing. Even when the Sun finally exhausts its fuel, we do not expect it to become a black hole. It should slowly collapse to a compact form called a *white dwarf*, roughly the size of the Earth.

A white dwarf can be no more than about 30 percent heavier than our Sun. Beyond that, a star will collapse until it becomes a neutron star, essentially one huge atomic nucleus some tens of miles in diameter. This will spin rapidly, throwing beacons of electromagnetic energy across the sky like a lighthouse. If the Earth happens to be in the path of one of these beams, which can consist of anything from radio waves to x-rays, we detect signals that steadily repeat like the ticking of a high-quality clock. A signal like this is known as a *pulsar.*

The first few pulsars were discovered in 1967 by Jocelyn Bell, then a graduate student working with the radio telescope at Jodrell Bank, England. Since then, hundreds more have been found. Bell's first hopeful guess as to what she had discovered was extraterrestrial life, so she fondly referred to the sources as LGMs, for "little green men."

A pulsar can serve as a very precise clock located outside the Solar System, and this provides yet another test of the relativity of time. The Earth's orbit is not a circle, and in early January it comes nearest to the Sun. This slows down earth clocks in two ways: the Earth is moving at its fastest, and it is at its lowest altitude in the Sun's gravitational field. The gravitational effect is twice as big as the one due to the Earth's motion. This discrepancy has been verified by comparing atomic clocks on Earth to one particularly accurate pulsar "clock" in the sky.

The formation of a neutron star releases, for a few seconds, as much energy as is produced by all the stars in our galaxy. This blows away the outer layers of the star to form a brightly glowing, expanding cloud called a *super-nova.* The Crab Nebula, the debris of a supernova recorded by Chinese astronomers in 1054, has a pulsar at its center, as do the residues of the supernovas seen by Tycho and Kepler. The supernova in the Large Magellanic Cloud in January 1987 gave the telltale signal of the formation of a neutron star, a titanic burst of subatomic particles called *neutrinos*, which will be discussed further in Chapter 19.

At somewhere above 1.5 times the mass of the Sun, a neutron star can no longer fight gravity and collapse must continue to a black hole. Given the small margin of additional mass required, many supernovas must ultimately take this route.

For obvious reasons, however, a black hole is hard to detect. Our best bet is to catch one that is sucking up matter at a substantial rate. This can happen if the black hole has a nearby binary partner that is a normal star. The black hole draws in hot gases from its companion's atmosphere. As they fall, the tremendous acceleration makes the gas radiate light; the higher the acceleration, the greater the frequency. A black hole has strong enough gravity to generate x-rays.

X-rays can be observed only from outside the atmosphere with orbiting telescopes. These have discovered many binary sources with heavy unseen companions, which must be either neutron stars or black holes. From its effect on the motion of its visible partner, the mass of the invisible partner can be calculated. In several cases, the mass is well above the theoretical limit for a neutron star. So it seems likely that these are black holes.

Additional clues come from a number of so-called "active" galaxies that pour out stupendous amounts of radiation from dense concentrations of matter at their centers. The only known way to power them would be a giant black hole with a mass of a hundred million Suns. Our own galaxy is tamer, but our view of its center is obscured by clouds of dust. Observers peering through this haze claim to see hints of the presence of more modest black holes.

So it is probably a safe bet that black holes are rather common. But given what happens to time in their vicinity, they can hardly be described as common*place*.

The ratio of time at the boundary of a black hole to that on the outside actually becomes *infinite!* The reader is urged to get a firm grip on something solid while contemplating the implications of that little tidbit. As seen from the outside, the black hole *never quite forms!* The last little bit of mass that would push it over the limit halts at the boundary. For practical purposes, however, it gets very close to the boundary in a few of thousandths of a second, so the "almost" black hole behaves just like a fully formed one.

But the point of view of an observer falling into the black hole is the real jolt. At the moment the boundary is crossed, the outside universe speeds up, flits through its entire history, and is snuffed out! As for the fate of our friend the observer, we must leave it to the imagination of the science fiction writers.

NOT WITH A WHIMPER

You may find it surprising that the universe as a whole is one of the easier problems in space-time geometry, but the explanation is simple: the universe is so huge that the individual stars or even galaxies are of no more consequence than atoms. We can take the distribution of matter as smooth and continuous.

When Einstein solved the problem in 1915, he got a big surprise; the universe cannot be static. It must either expand or contract. Since the idea of an eternal, unchanging universe was at that time firmly implanted in the human mind, Einstein did not take the result seriously but added a fudge factor, the so-called *cosmological term*, to make a static universe work. He was later to call this "the biggest mistake of my life," for in 1927 the American astronomer Edwin Hubble demonstrated that the universe really is expanding!

What Hubble showed was that all remote galaxies are moving away from our own, as if it were some cosmic untouchable. The farther away they are, the faster they flee. General relativity tells us that matter and space-time are inseparable. The galaxies are not simply spreading out in infinite, empty space, but space itself is finite in size, and it is growing.

The balloon in Figure 12-5 illustrates how this can happen. Let the galaxies be spots on the balloon. As it inflates, each spot moves farther from its neighbors. The farther apart they are, the faster they move away.

This is of course a two-dimensional example. Our space has three dimensions. The fourth dimension, time, is the expansion itself. And there being no additional space dimension, there is no "inside" to our cosmic balloon.

Because of gravitational attractions, the expansion is continually slowing down. Will it eventually go into reverse and become a contraction? The answer is that we don't know, because we are not sure how much mass there is in the universe. The mass we do see is not enough to turn the trick, by a wide margin. But there may well be more mass out there unseen, waiting to be discovered. The motions of stars within galaxies show gravitational effects that point to a "halo" of invisible mass many times greater than that which is seen. Black holes and unknown forms of radiation could all contribute. Our estimate of the total mass of the universe can only go up. Still, at the present our most optimistic estimates fall short of what is needed to reverse expansion by about a factor of 10.

FIGURE 12-5.

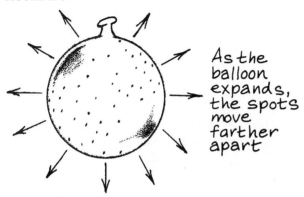

As the balloon expands, the spots move farther apart

Nonetheless, many scientists continue to believe that the expansion will eventually reverse, largely because the alternative—to blow away into nothingness—is far less appealing.

Projecting the expansion 12 billion years backward in time, we see the birth of our universe in a titanic explosion, called the *big bang* by cosmologists, specialists in the esoteric art of contemplating the origins of absolutely everything.

At this point it is important to insert a *caveat*. Most scientific disciplines shun speculation that goes much beyond what can be tested by observation. But solid observations relevant to the early universe are few and far between, and we are too human to avoid speculating about anything so inherently fascinating. So in cosmology, normal scientific caution goes out the window and fanciful theories are up for grabs.

Anything that rapidly expands must cool, and our universe is no exception. It must have been born incredibly hot. Under such conditions, matter conducts electricity freely and is opaque to all forms of light. The universe remained in this state for a few hundred thousand years. When it became cool enough, light and matter parted company in a brilliant flash that should still be rattling around the universe today. It has been seen, and looks exactly as predicted back in 1948 by George Gamow, a Russian-born American physicist of whom we shall hear more in Chapter 17.

Light, too, cools as it expands. What this means is that it shifts to longer wavelengths. Today, this "echo" of the big bang has been stretched into radio and infrared waves ranging in wavelength from micrometers to millimeters. They are known as the "2.8 K background," because their effective temperature is 2.8 kelvins (K), which means 2.8 Celsius degrees above absolute zero. The big bang theory is quite explicit about this temperature; how it is measured will be explained in Chapter 15. But the fact that the theory predicts it so precisely is the same sort of "crucial link" that the moon's acceleration was for Newton's gravitational theory.

Interestingly, the 2.8 K background *does* define a reference frame that one can regard as *somewhat* privileged. While this radiation moves at the same speed in all reference frames, there is only one frame in which there are equal amounts of it moving in all directions, for a total momentum of zero. If we are moving with respect to that reference frame, we run into a bit more of the radiation in the direction in which we are moving. This effect has been detected, and shows that the Sun and Earth are moving at 600 km/s in that frame. If it were privileged in Michelson's sense, he would have seen a dramatic effect—a shift of *forty wavelengths!*

Another important feature of the 2.8 K background is that it is not entirely equal in all directions. Precise measurements conducted since 1992 have revealed tiny irregularities at a level of about one part in 100,000. This, as it turns out, is very welcome news, for it shows that the matter that emitted this radiation was not quite uniformly distributed in space.

In a uniform cloud of matter, gravity would have pulled equally in all directions. There would have been no way for matter to form concentrated

George Gamow.
(Courtesy of American Institute of Physics.)

clumps that would eventually become galaxies and stars. For this process to get started there had to be "seeds" of higher density toward which neighboring matter would be attracted.

The speed with which galaxies form depends on the total amount of mass in the universe. The more there is, the stronger the gravitational forces, and the more rapidly galaxies coalesce. The first galaxies formed within a few billion years of the big bang. For this to have happened starting from the tiny observed irregularities in the 2.8 K background, there had to be considerably more matter than we have seen. Once again, here is another hint to encourage those who want to believe that the expansion of the universe will ultimately reverse itself.

But how about the very instant of creation, when our universe burst forth from what may have been a geometric point? And what will happen if the expansion does reverse, and the whole mess rushes back together in one cataclysmic gravitational collapse, the "big crunch"? We can do little more than guess. But our guesses will be far more intelligent after we learn something about the quantum theory, the physics of the microworld. So we must now reluctantly drop this fascinating topic, and pick it up again in the final chapter.

THE EINSTEIN CULT

Like Newton before him, Einstein lived to be a legend in his own time, the only twentieth-century scientist whose name and face bring instant recognition. One of the reasons for this, as it had been for Newton, was the urgent need of the scientific community for a hero to put on public display. Einstein was ideally cast in the role. With his modesty, his rumpled clothing, his bemused expression, and above all his warm humanity, he personified the brighter side of science.

Fame came to Einstein in 1919, in the wake of the eclipse expedition. Eddington was a master at milking a discovery for the maximum of publicity. In this case, he also had a political ax to grind. World War I had ripped the international fabric of science asunder. In the wake of the carnage wrought by modern weapons, science and technology stood tarnished with blood. Science needed above all to show the world its humanist face and proclaim anew a tradition that stood above the petty hatreds that set nation against nation.

The eclipse measurements were announced in a fashion that can only be described as high stagecraft. At a large scientific meeting at Cambridge University, a packed house was told that Einstein had been proved right. Overlooking the lectern was a bust of Newton, and Eddington dramatically turned its face to the wall. Over the next few days, the world's newspapers proclaimed, in sensational headlines, a revolution in science. "Stars All Askew in the Heavens . . . Nothing Where it Seems to Be . . . Even the Multiplication Tables in Doubt . . . But No Worry, Dr. Einstein Understands All."

Eddington proudly boasted that British scientists had traveled thousands of miles to test the ideas of a German colleague, in an expedition launched while the two nations were still technically at war. It was known that Einstein had lived under somewhat of a cloud in Germany for opposing the war, as had Eddington, who was a Quaker, in Britain. The world was ready for heroes like that. Moreover, Einstein *looked* the part of the gentle "sage on the mountain," deciphering the mystery of the heavens.

However well Einstein may have suited the role, it most emphatically did not suit him. He had done his best work in obscurity and relished his solitude. The perquisites of celebrity had little appeal to his simple tastes. Near the end of a triumphal world tour in 1921, the year of his Nobel Prize, Einstein was conveyed across Spain in the king's own private railway car. He found the experience so distasteful that he completed the journey to Berlin by third class.

Liberal social thinkers climbed on the Einstein bandwagon, professing to see in his theory a reflection of their own principles of moral and cultural relativism. How could anyone still believe in absolutes in morals, customs, or politics, when not even the perfect world of physics had room for them? Einstein demurred: he had been misunderstood. Relativity had not abandoned absolute truth, only the false absolutes of space and time. Once he accepted Minkowski's four-dimensional invariants, Einstein even made a feeble try to change the name of relativity to *invariantstheorie*.

Within his own profession, Einstein was to remain a loner, aloof from the mainstream. The better physicists became adept at relativistic calculations, at least with the special theory, and most admired its beauty. But Einstein, however celebrated and respected, never took a leadership role in the physics community.

This was due in part to Einstein's career, much of which was spent in positions that brought him little or no contact with students. His solitary work habits were another barrier, preventing younger colleagues from mastering his style. And the theories themselves were partly to blame. Special relativity was complete within a few years after 1905. There was no need for a large school of specialists to grind out predictions or for a corps of experimenters to test them. The general theory was even worse, since until after his death it was largely inaccessible to experiment.

But most of all, Einstein was isolated by his own personal worldview, which swam against the intellectual tides of his day. For Albert Einstein was an uncompromising rationalist, with a deep faith in the underlying logic of the universe. He often personified this faith with the name of God, or "the Old One," though he steadfastly refused to practice any organized religion.

In his youth, Einstein had been strongly influenced by the positivist philosopher-physicist Ernst Mach, whose insistence that physics be firmly rooted in a critical examination of the process of observation was an obvious guidepost in the development of relativity. But as he matured, Einstein came to view purely empirical science as "Mach's little horse," a creature "who can only exterminate harmful vermin" but "who cannot give birth to anything living." In this attitude Einstein revealed a spirit even more Platonist than that which animated Galileo.

In Einstein's universe there was no room for the arbitrary, irreducible fact. There must be no limit to the power of the human mind to reveal why the universe must be exactly as it is. Niels Bohr and his disciples, on whose work we shall soon focus, rejected this view utterly, and even professed to have shown there are limits to the power of reason to fully comprehend nature.

But Einstein clung tenaciously to his lonely stand. In his later years, he summarized his approach to nature in one phrase: "What really interests me is whether God had any choice in the creation of the world."

Summary

General relativity grows out of special relativity by adding the *Principle of Equivalence,* which formalizes a connection between gravity and accelerated reference frames. This leads to a reformulation of mechanics in which gravity is accounted for through curved geometry for space-time, and there is no need for force. This leads to a dependence of time on *altitude* in a gravitational field. It also predicts deflection of light by the Sun's gravity that is twice that calculated from Newtonian gravity. The resolution of this test in Einstein's favor through measurements made during an eclipse and made Einstein a celebrity. The theory also implies a nonstatic universe, confirmed by later observations, and allows for extreme distortions of space-time now known as *black holes.* Though general relativity is now the basis for cosmology, it remains an incomplete framework until forces other than gravity are brought into the geometry.

CHAPTER 13

The Atom Returns

The telescope at one end of his beat, And at the other end the microscope,
Two instruments of nearly equal hope . . .

—ROBERT FROST, The Bear

It was his early work on relativity that marked Einstein as a genius in the eyes of a few first-class theorists, such as Planck and Lorentz. But the questions it addressed were not, at that time, uppermost in the minds of most physical scientists, who were preoccupied with what was considered a more urgent crisis—the debate over whether atoms were "real." Einstein played a role in bringing this debate to a decisive conclusion through work that displayed him as a solid professional capable of advancing and influencing the work of others, rather than as simply a creative loner with his own private agenda. This contribution helped to solidify his position in the broader community of science.

To put this dispute in context, in this chapter we recapitulate the story of how the atom reentered respectable scientific thought during the nineteenth century.

THE ANCIENT ORIGINS OF ATOMISM

It is mandatory to preface any discussion of atoms by paying homage to Democritus of Abdera, a philosopher of the fifth century B.C. Though he is generally classified as an early or "pre-Socratic" philosopher, Democritus actually outlived Socrates and is known to have written on nearly as broad a range of topics as Aristotle. Unfortunately, only fragments of his writings survived, and we know of his ideas largely through the admiring commentaries of Aristotle and others. Through these, he has become celebrated as the father of the atom. It was he who coined its name, the Greek word for "uncuttable."

Plato embraced the atom, as did Epicurus, founder of another influential school of Greek philosophy. The word "school" can here be taken in its customary sense: these philosophers were professional educators who founded institutions that endured for centuries, training the elite of Greek and later Roman youth. The Roman poet/philosopher Lucretius immortalized Democritean atomism in *De Rerum Natura*, a widely read and influential work.

Thus in the ancient world, the atom was widely (but by no means universally!) accepted.

The idea of the atom was appealing in several ways. Chief among these, it explained how matter could be transformed, but never created or destroyed. Nothing comes from nothing, and nothing disappears without a trace. Atoms may be rearranged, but they endure forever.

To the scientist confronted by the sheer variety and complexity of nature, atomism promised hope. Perhaps nature on the small scale might be far tidier than the world of everyday experience. Underlying all this confusion might be a level of reality of stark simplicity, with the turmoil we perceive representing only the nearly infinite variety of arrangements among a myriad of small parts.

Nonetheless, from the outset atomism had its critics. One of the most distinguished was Anaxagoras, an older contemporary of Democritus who asked what atoms might be made of. If they were simply little blocks of some continuous substance, it would be that substance that would be of scientific interest: a brick wall derives most of its properties from the fired clay of which bricks are made, rather than the size or shape of the bricks themselves. The alternative was for atoms to have smaller parts, which would in turn have even smaller parts, for an infinite chain of "seeds within seeds," leaving simplicity forever an illusion to be sought in the next layer of the cosmic onion.

Viewed this way, atoms looked far less appealing. And modern scientists, committed as they are to the atom, can only admit that so far the history of atomism has followed Anaxagoras' scenario faithfully.

It was not arguments such as these but rather the rise of Christianity that brought atomism into disfavor. It was condemned for its association with the philosophy of Epicurus and Lucretius, a tolerant and skeptical worldview incompatible with the zealotry of the Church fathers. Furthermore, some Christian theologians saw in the enduring unalterability of atoms a denial of *transubstantiation,* the miracle of the Eucharist, in which bread and wine literally become the body and blood of Christ.

Thus, throughout the Middle Ages atomism was regarded in the Christian world as bordering on heresy. But it was never completely suppressed, because it explained in a qualitatively satisfying way many simple properties of matter. Its most notable success was the way in which it accounted for the properties of the three phases of matter—gas, liquid, and solid. The rigidity of a solid shows that its atoms are hooked firmly together. In a liquid they are still in contact but free to move around, which is why fluids settle into the shapes of their containers and yet remain as difficult to compress as solids. A gas, finally, can expand to fill any container because its atoms are widely separated and are moving rapidly. This picture survives to this day.

Galileo, Descartes, and Newton were atomists, though they did little to extend the theory. It remained for the chemists of the early nineteenth century to find the first solid empirical support for atomism. Without stretching the point too far, it is fair to say that in 1800 the atomic theory was something physicists believed but couldn't prove, while the chemists were proving it but didn't believe it. Thus, at this point a brief digression into chemistry is in order.

THE BIRTH OF MODERN CHEMISTRY

The latter half of the eighteenth century had been to chemistry what Galileo's time had been for physics. The outstanding achievement had been to put chemistry on a sound quantitative basis. This approach helped lead to a number of important discoveries. One of the most significant was clarifying the distinction between a true chemical reaction and a mere process of mixing. This distinction had been dimly perceived before; mixtures displayed properties that were a blend of those of their components, in a manner that depended on their relative proportions. A chemical reaction, however, might produce a substance totally unlike the materials that went into its formation.

Common water, for example, arises from the union of the gases oxygen and hydrogen. Similarly, the puttylike metal sodium reacts with the green gas chlorine to form ordinary table salt. But at times the basis for the distinction seemed hazy, until precise weighing revealed the key. Mixtures could be formed in any desired proportions, but chemical reactions must follow an exact recipe. The constituents had to be present in some exact proportion of weights. If too much of one of them was present, some would be left over after the reaction. Finally, the founders of modern chemistry had clarified the distinction between *elements,* which could not be broken down into other substances, and *compounds,* which could.

The whole picture was terribly inviting to an atomist. Elements must represent the different kinds of atoms. Compounds are then substances formed by attaching atoms of different elements together. Mixtures arise from the free mingling of independent atoms without any ties between them. But atomism, and indeed the whole intellectual style of imaginary model building that lay behind it, was mainly the province of physicists. One such, the Italian Amedeo Avogadro, pushed the atomic idea in chemistry before 1800.

Throughout its history, chemistry has tended to be a far more conservative science than physics, sticking close to its empirical roots and practical techniques and disdaining abstraction and speculation. The chemists paid little attention to atomism until one of their own number, the English chemist Thomas Dalton, brought it forcefully to their attention by showing that an atomic structure to matter could explain the peculiar regularities that kept popping up in the recipes for chemical compounds.

This regularity was expressed in the law of constant proportions. Stated crudely, it indicated that the amounts of an element that entered into forming all its compounds were related. Hydrogen, for example, was always vastly outweighed by its partner when entering into combination, while lead always dominated its compounds. In more exact terms, it was found that each element seemed to have a characteristic *equivalent weight*. Hydrogen was the lightest and could be taken as the starting point on the scale. Oxygen was eight times heavier, sodium twenty-three times, chlorine thirty-five, and so on. All recipes for compounds could be formed from these equivalent weights.

In the first decade of the nineteenth century, Dalton pointed out that the whole scheme could be simply understood by taking the equivalent weights to represent the relative weights of the atoms of the elements. Then the recipe for common salt, 23 parts sodium to 35 parts chlorine, merely represented the fact that chlorine atoms were 35/23 as heavy as sodium atoms, and salt was formed by joining each sodium atom to one chlorine atom. Such a combination Dalton called a *molecule*. The molecule is the smallest constituent of a chemical compound, just as an atom represents the smallest unit of an element.

Still, a lot of facts remained unexplained. Some elements seemed to have more than one equivalent weight. It gradually became clear that the pairing off of elements into two-atom molecules proposed by Dalton was too simple; some molecules must contain three or more atoms. For example, oxygen atoms prove to be sixteen, not eight times as heavy as hydrogen. The proportion 8:1 of oxygen to hydrogen in the recipe for water reflects the fact that two hydrogen atoms join each oxygen atom when a molecule of water is formed.

As is usual in such situations, there was a certain amount of bad data in circulation to confuse the issue further. It took fifty years to untangle the mass of chemical data, but in 1858 a young Sicilian chemist, Stanislao Cannizzaro, published a compendious review that finally established the correct relative weights of the atoms of the better-known elements and gave the atomic composition of their known compounds. The atomic theory has been the foundation stone of chemistry ever since. Nearly all chemists spoke of atoms, though many still refused to accept them as real. Cannizzaro, however, was revolutionary by temperament: he went on to join Giuseppe Garibaldi in the *Risorgimento*, the revolution that gave birth to modern Italy.

THE PHYSICISTS PICK UP THE BALL

The success of atomism in chemistry was bound to encourage the physicists in their natural predilection for the theory. Old ideas were resurrected, cloaked in a new mantle of respectability. One important idea dated back to Newton's contemporary Robert Hooke, a confirmed atomist. Hooke speculated that the outward pressure exerted by a gas on the walls of its container might originate in a hail of atoms. Each atom exerts a force on the wall when it hits, and there are so many such impacts that the result seems a steady outward push.

Hooke found support for his view in the work of Robert Boyle, performed a generation before Newton and Hooke. Boyle found that if a gas is compressed in a closed container, as shown in Figure 13-1, the pressure on the walls varies inversely with the volume. If the piston is pushed in far enough to reduce the volume by half, the pressure of the gas will double.

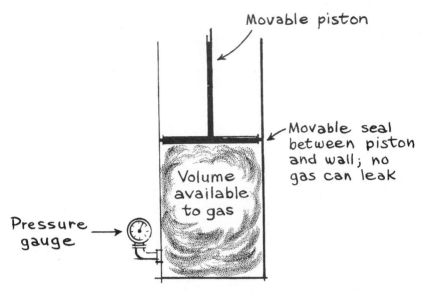

FIGURE 13-1.

This effect is quite easy to understand in atomic terms. With half the volume to roam in, the atoms are packed in closer, so there are twice as many of them in any region of the cylinder. If their motion is unaltered by this crowding they strike the walls with the same impact, but twice as many impacts take place, doubling the pressure. Later experiments on the effects of heat on gases gave this idea further support.

Even before the days of Galileo the notion that heat might represent some form of microscopic motion enjoyed some vogue. Francis Bacon, the fifteenth-century English philosopher, accepted the idea. The work of Joule on the conversion of motion to heat made it even more appealing.

Heating a gas in a sealed container always causes a rise in pressure. Studies on the behavior of gases heated in closed containers showed that the pressure seemed to rise and fall linearly with temperature, as shown in Figure 13-2. No matter what gas was used, the pressure always behaved as if it would reach zero at the same temperature, −273° C. All these studies, of course, were carried out at much higher temperatures: it was only by extending the straight line graph to lower temperatures that this feature was discovered. Atomists christened this temperature "absolute zero" and argued that it represented the cessation of all atomic motion.

In 1847, Rudolf Clausius showed that one could account completely for the behavior of gases by assuming that the "absolute" temperature, i.e., that measured from absolute zero, is simply a measure of the kinetic energy of molecules. If pressure is due to a hail of molecules, speeding them up raises the pressure in two ways. First, the rate of collisions with the walls of the container increases. Second, the force due to each molecular impact is increased.

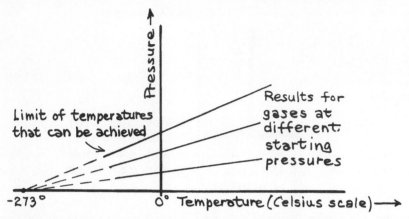

FIGURE 13-2. Pressure changes due to heating and cooling gases.

Since the first effect is proportional to velocity, and the second to momentum, the pressure must be proportional to mass times velocity squared, and thus to the kinetic energy.

It soon became clear that it was unreasonable to assume that all molecules in a gas were moving at the same speed. The kinetic energy Clausius spoke of was simply the *average* value. The details of the picture were filled in over the next few decades by a number of gifted contributors, including James Clerk Maxwell, the Austrian theorist Ludwig Boltzmann, and an American named Willard Gibbs. The theory in its most sophisticated form is called *statistical mechanics*, and was completed by the end of the nineteenth century. As its name suggests, the theory does not attempt to describe the motions of individual atoms, but treats them statistically, as demographers do census data.

HOW BIG IS AN ATOM?

One vexing problem remained to cloud the success of atomism: neither physicists with their gas laws nor chemists with their reaction recipes could say with any certainty what the actual masses of atoms were. Pressure depends on the sum of molecular masses, not the mass of an individual molecule. Chemistry showed that an oxygen atom was sixteen times heavier than one of hydrogen, but no one knew how much either actually weighed.

At about the same time, the microscope reached its theoretical limit of perfection. Objects a bit smaller than a wavelength of light could be seen, but still the atom remained invisible. On the philosophical side, the movement called *positivism* was at its zenith, and exponents such as Ernst Mach were urging that everything unobservable be expunged from science. In these circumstances, the natural conservatism of chemists reasserted itself. Wilhelm Östwald, the father of modern physical chemistry, spearheaded an attack on the atomic theory.

The successes of atomism, some of which were Östwald's own work, could hardly be ignored. He was willing to retain the atom as a "heuristic" concept, one used as an aid to understanding but not taken literally, like the classification schemes employed in biology. Ultimately, he hoped, atomism could be replaced by an extension of the energy concept. Though this remained a minority view, its adherents were prestigious and their arguments could not be dismissed out of hand.

In the eyes of most scientists, the dispute hinged on measuring "Avogadro's number" N, the number of atomic mass units in a gram. A variety of techniques had been devised, but none were yet terribly reliable. In the 1890s, the highest estimates of N were more than 100 times the lowest values. If several methods could be perfected, and gave the same result, the atom was on solid ground. If not, Östwald might well be right.

The debate was raging in 1905, Einstein's "miracle year," and despite his other interests he did not ignore it. Statistical mechanics had in fact been the focus of his scientific work prior to that year. Thus three of the six papers (five published) he wrote in 1905 concerned the measurement of N. One was his doctoral thesis (unpublished) for the University of Zurich, which proposed studies of the diffusion of large molecules such as sugar in solutions. What earned him more attention were two published papers on the phenomenon of *Brownian motion*.

In 1827 the English botanist Robert Brown had used a microscope to study pollen grains suspended in water. He found that the grains refused to sit still, hopping about in a jerky erratic fashion. Convinced in advance that pollen consisted of inert spores with no means of locomotion, Brown showed that similar motions took place when pollen was replaced by similar-sized particles of dust or soot.

A few of Brown's contemporaries suggested that the motions might be due to random imbalances in the molecular impacts on opposite sides of a pollen grain. But the topic was soon forgotten, and late-nineteenth-century physicists and chemists had to rediscover Brown's work on their own. By 1905, many had realized that quantitative studies of this effect might be a good way to measure N, and several attempts had been made.

In the first and fourth of his 1905 papers, Einstein outlined a particularly simple experiment. Prepare a suspension of spherical particles of equal size. Under a microscope, focus on one particle at a time and record how long it takes to wander from the center to the edge of a circle of known radius. Since this is a random process, the measurement must be repeated many times and the results averaged, but with each successive measurement the average becomes more reliable.

The advantages of this method impressed a young but established French experimenter named Jean Perrin. He had been working on the problem from a somewhat different angle, but quickly adopted Einstein's approach. This endorsement, above all, solidified Einstein's reputation.

By 1909, measurements of N by this and other techniques had narrowed to the range of 6 to 9 times 10^{23}.* Östwald graciously conceded defeat, proclaimed the Brownian motion the decisive factor in his conversion to atomism, and praised the contributions of Einstein to the debate. He was acknowledging more than just a scientific misjudgment. In 1900, he had turned down an application for the position of research assistant from one A. Einstein of Zurich!

WHAT'S "INSIDE" AN ATOM?

The natural next step was for a few bold souls to speculate on what an atom might really look like. Though atomists expected atoms to be simple and nearly structureless, by 1900 there was plenty of evidence to the contrary. First and foremost, some means had to be found for hooking atoms together into molecules. Second, the existence of trends and similarities in chemical properties indicated by the periodic table of the elements was strongly suggestive of underlying structure.

Still, "atom" meant *indivisible,* and it seemed futile to speculate about dividing the indivisible until two sensational discoveries just before the turn of the century. These were the discovery of radioactivity by Henri Becquerel in 1896 and of the electron by J. J. Thomson one year later.

Becquerel's discovery had the more sensational popular impact. One of the marvels of the nineties had been the discovery of x-rays. The practical implications of a form of "light" that could penetrate opaque objects titillated the late Victorian public and led to sensationalized and amusing newspaper articles. Following a hunch, Becquerel tried to find a substance that would give off x-rays when placed in ordinary light. Instead, he found that pitchblende, an ore containing uranium, did indeed give off radiation, but it was not as penetrating as x-rays and seemed to arise spontaneously not only in the absence of light, but in fact oblivious to all outside influences. No amount of heating, treating, or cajoling could change the inborn rate at which a radioactive substance gave off rays.

Becquerel's student Marie Curie discovered that pitchblende contained not one but several radioactive elements, the most powerful of which she gave the name *radium*. Every few days it radiated as much energy as would be released by an equal weight of the most powerful explosives, but continued to radiate undiminished!

The practical implications of this were realized immediately. Within a few years, the novelists Anatole France and H. G. Wells were writing fantasies about "atomic bombs," though thirty years would elapse before the discovery of nuclear fission made these a reality. What was the source of this fearsome energy?

---

*The accepted modern value is 6.02×10^{23}.

The chemist Frederick Soddy and the physicist Ernest Rutherford (of whom we shall hear a great deal more in the next chapter) guessed that the energy came from *within* the atom, and established that radiation was accompanied by the transformation of one kind of atom into another. When Soddy proposed the word "transmutation" to describe this process, Rutherford was aghast—"They'll hang us for alchemists!" Nineteenth-century chemists had proudly boasted that they had shown the futility of the alchemist's quest to turn base metals into gold. Atoms were immutable, and that was that.

This discovery made probing within the atom not only respectable but imperative. Anything that could spontaneously change in such a dramatic fashion must have internal workings of quite a complex order. And one of its parts—the *electron*—had just been discovered.

IONS AND CATHODE RAYS

Back in the 1830s Michael Faraday had studied the conduction of electricity in liquids. There the flow of electricity is usually accompanied by an actual movement of matter to the electrodes through which the current enters and leaves the liquid. For example, the passage of electric current through water results in the liberation of hydrogen at one electrode and oxygen at the other, a phenomenon known as *electrolysis*.

Faraday found that the amount of an element that arrived at an electrode was proportional to the total electric charge and to the equivalent weight. To anyone who believed in an atomic picture of matter, the interpretation of this law was obvious: all one needed was that electricity be transported in the form of an electric charge on the atoms. For some unknown reason all atoms regardless of type carried the same unit of electric charge or a simple multiple of that unit. Faraday called his charged atoms *ions,* a Greek word meaning "wanderer" and thus a fit companion for the term *atom* itself.

J. J. Thomson had followed in the giant footsteps of Maxwell as director of the Cavendish Laboratory. In keeping with that institution's ties to the mushrooming electrical industry, he was studying the conduction of electricity in tubes filled with gases at low pressure, research that eventually led to neon signs and fluorescent lights. Others working on this problem had discovered *cathode rays,* so named because they moved from the negative electrode, or *cathode,* to the positive one, the *anode.* These behaved the same whatever gas was used to fill the tube. Thus it was unlikely that they were Faraday's ions, but what else could they be?

Thomson reasoned that if the gas were sufficiently rarefied, cathode rays could travel great distances without colliding with an atom. In unimpeded flight, they could be probed by seeing how they react to electric and magnetic forces.

Thomson suspected he was dealing with some new sort of particle, carrying a negative electric charge. The force on an object in an electric or magnetic field is proportional to its electric charge, and the acceleration produced by

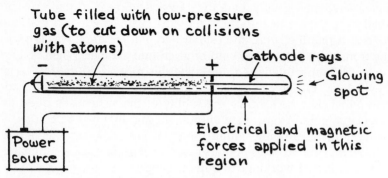

FIGURE 13-3. Apparatus for the discovery of the electron.

that force depends on the object's mass. Thus, any electromagnetic experiment measures the ratio of charge to mass. For ions, of course, Faraday had measured that ratio years before, so Thomson would have a standard against which to compare his measurement.

When a hole is pierced through the anode, a beam of cathode rays passes through and leaves a telltale glowing spot where it strikes the walls of the tube, as shown in Figure 13-3. Thompson found that moderate magnetic fields easily deflected the beam, but it took enormous electric fields to budge it at all. This suggested that the cathode rays were moving very fast. The magnetic force on a charged particle is proportional not only to its electric charge but also to its speed, whereas electric force depends on the charge alone. Thus magnetic fields are far more effective in deflecting fast-moving particles than electric fields, which do not benefit from the increased velocity.

Comparing the relative strengths of the electric and magnetic fields required to produce equal deflections of the cathode-ray particles enabled Thomson to calculate their velocity.* They proved to be fast, indeed—more than one-tenth the speed of light, an unprecedented speed for material objects!

Once the speed was known, Thomson could use the deflection to calculate the acceleration, and thus the ratio of charge-to-mass. The result was surprising: nearly 2000 times higher than the charge to mass ratio of hydrogen, the lightest known ion. Since the ratio proved independent of the gas in the tube, Thomson in 1897 announced the discovery of a new form of matter, which was christened the *electron*.

Either Thomson's electrons carried an enormous charge compared with ions, or they were far lighter. But if they originated from atoms in the gas, which then became positive ions, how could one get the charges to balance? This conjecture, along with their speed and penetrating power, pointed to a startling conclusion: cathode rays consisted of streams of particles that carried the same charge as ions, but were nearly 2000 times lighter than a hydrogen ion.

---

*In a magnetic field of strength B the force on a particle of charge q and velocity v is qvB, while in an electric field E it is qE. Thus if the forces are equal, $v = E/B$.

THE ELECTRON AND THE ATOM

Thomson's electron was immediately hailed by some as the "atom of electricity," the solution to the mystery of the nature of electricity. The existence of this unsuspected light object seemed to explain the great mobility of electricity. But the true significance of his discovery was by no means lost on Thomson. His electron must be *a component of the atom itself.*

Where lesser minds saw the solution to an old mystery, Thomson saw the opening of a new adventure. A potential building block of the atom had been uncovered, and the rush was on. Led by Thomson himself, the more daring citizens of the world of physics began inventing hypothetical models of the atom, not as a mere pastime or for purposes of illustration, as a few years earlier, but in dead earnest. A few clever experiments might tell physicists what really went on in an atom!

In their bold optimism, these early speculators on atomic structure little realized that they had opened a Pandora's box, and that the demons that would emerge from it had no place in the tidy world of classical physics.

Summary

Though widely accepted in antiquity, the atomic hypothesis was rejected in the Middle Ages for theological reasons. It reentered mainstream science in the nineteenth century, first in chemistry and later in physics, where a statistical treatment of atomic motions was able to explain many kinds of transformation of energy. But as long as atoms remained unobservable and their true size was unknown, many good scientists refused to accept their reality. Developments in the first decade of the twentieth century ended these doubts, and Einstein's analysis of Brownian motion played a significant role. By then three discoveries—x-rays, radioactivity, and the electron—had paved the way for serious attempts to determine the internal structure of atoms.

CHAPTER 14

Rutherford Probes the Atom

The universe is not only queerer than we imagine, it is queerer than we can imagine.

—J. B. S. HALDANE

Imagine a group of proud and inventive people quarreling over the contents of a sealed box and you have a pretty good picture of the mood of the early thinking on atomic structure. Electrons were clearly an important constituent, but how were they arranged? And given that they weighed so little, and only carried negative electricity, there was a great deal of mass and positive electric charge to be accounted for. Most of the speculation centered on one of two models: the *planetary* and the *plum-pudding* schemes.

Given the similarity between Coulomb's Law of Electrical Force and Newton's Law of Gravitation, an atom that resembled the Solar System, with a massive, electrically positive "Sun" and negative electrons swinging around it in Keplerian orbits was too pretty an analogy to pass up. Furthermore, it placed some powerful computational tools developed over two centuries of study of planetary orbits at the disposal of the theorist.

But the opposing camp had as its prime asset the formidable authority of its founder, none other than the illustrious J. J. Thomson himself, who became one of the first Nobel laureates in physics for his discovery of the electron. He proposed a sphere of positive charge in which the electrons were embedded, as shown in Figure 14-1; the descriptive term *plum pudding* was his own choice.

LIGHT EMISSION IS THE TEST

Thomson's model would have been regarded as idle speculation had there not been a body of data crying for explanation by an electrical model of the atom. These were the data on the emission of light by atoms.

Light from a source containing a single element in a gaseous state, such as a neon sign or a mercury lamp, always has a characteristic color. When this light is broken up into its component colors by a prism, a striking result is obtained. Instead of a continuous spectrum (rainbow), which occurs when a solid or liquid is heated to glowing, one finds the light is composed of a few very pure, sharply defined colors.

With two
electrons

With four electrons
(they are at the
corners of a
regular pyramid)

FIGURE 14-1. Plum-pudding atoms.

The best way to observe this is to pass the light through a thin slit, as illustrated in Figure 14-2. On a viewing screen or photographic plate it will form a pattern of thin lines, each of a different color. For this reason, this type of pattern is called a *line spectrum*. (See Figure 14-3 for a portion of the line spectrum of helium.) For most elements, only a few lines are bright enough to be seen with the naked eye. But if a long photographic exposure is made, many fainter lines appear; for some elements, hundreds have been cataloged.

Maxwell's electromagnetic theory allows for only one way to produce light of pure colors: somewhere, an electric charge must be going through a regular, periodic motion. The frequency of this motion determines the frequency of the light. This was one reason why Thomson's insistence that his electrons must be part of ordinary matter was so readily accepted; anything that emitted light had to be electrical in nature.

It was also clear that in a gas the light must be emitted by individual atoms. Not only are the atoms separated by many times their own size, but the oscillations of light waves have enormous frequencies. There are many oscillations in the time between encounters with another atom, and thus there is no way for several atoms to cooperate in such a rapid oscillation. Furthermore, the striking differences between the continuous spectrum of light emitted by densely

FIGURE 14-2. Apparatus for producing line spectra.

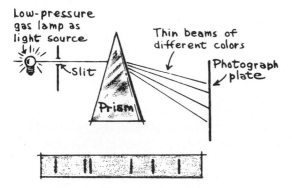

FIGURE 14-3. A portion of the line spectrum of helium.
(Mount Wilson and Palomar Observatories.)

packed solids or liquids and the line spectrum from rarefied gases gave support
to the notion that line spectra represent light from individual atoms.

Thomson's model gave a quite natural explanation for this light emission.
He imagined that his electron "plums" were able to move freely in their positively charged pudding, held in place by a delicate balance between their
attraction to the center of the positive charge and their mutual repulsion. A
single electron would rest at the center of a sphere, while three would form an
equilateral triangle and four a tetrahedron. If disturbed from these patterns by
a collision between atoms, they would oscillate around their normal positions,
just as a pendulum oscillates when disturbed from its equilibrium point.

Whenever a charge is accelerated, an electromagnetic field radiates out as
light, carrying away the energy of motion. The vibration must quickly die out.
The natural frequency of these oscillations set the frequency of the waves emitted. For atoms around the known atomic diameters of a few times 10^{-10} m,
they could be shown to be appropriate frequencies for visible light. This was a
very encouraging result: that the electron had just the right amount of charge
and mass to connect the size of the atom to the frequency of light had to be
more than a coincidence. The plum-pudding atomists were sure they must be
on the right track.

But the same numerical coincidence also encouraged the planetary enthu-
siasts. Orbits of atomic diameter gave the right frequencies of rotation to produce visible light. But this also proved the undoing of the model. Unlike
Thomson's atom, there was no natural way to stop the light emission or to
give it a constant natural frequency. An orbit could have any size and thus
could radiate light at any frequency. As the electron lost energy, its orbit would
gradually shrink. Spiraling in to its doom, the electron would gradually
increase its frequency of rotation. Thomson's electron oscillations kept the
same frequency as their amplitude decreased.

Even worse, simple calculations using Maxwell's laws showed that it took
no longer than a millionth of a second for the orbit to shrink to a tiny fraction
of its original diameter. The planetary atom was unstable and gave no natural
explanation of the line spectrum. Despite heroic and ingenious efforts to eliminate these faults, the model fell into disfavor.

Still, until the Thomson model could be shown to explain the observed
spectrum lines in all their quantitative detail, the field was open to all comers.
And the quantitative detail was immense. The wavelengths of spectral lines
are among the easiest physical quantities to measure to high precision, and
precise data had been piling up for decades, thanks in part to their practical

value. The set of spectrographic lines produced by an element is its finger-print, and spectrography was a marvelous tool for chemical analysis. One bright line could reveal a small trace of one element in the presence of another that had only faint lines near the same wavelength.

While the chemists were content to assemble their fingerprint file, physi-cists were doing much of the actual experimental work, because the tech-niques involved fell in the realm of optics. And the physicists could not resist the temptation to search for regularities in their data.

ORDER IN THE CONFUSION OF SPECTRAL LINES

For most elements, a list of frequencies of spectrum lines shows no more order than, for example, a list of social security numbers. The exceptions are those elements that fall in column 1 of the periodic table of the elements, hydrogen and the so-called alkali metals. A regular pattern in the hydrogen spectrum was first noticed in the 1880s by Johann Balmer, a teacher at a Swiss technical school. Today we express this regularity by the formula

$$\nu = \nu_0 \left(\frac{1}{n^2} - \frac{1}{m^2} \right)$$

where ν_0 is a constant and n and m are integers, with m greater than n. Every pair of integers gives a different spectrum line. It was soon found that the same formula very nearly held true for the alkalis.

The plum-pudding model offered no explanation for this formula, or for the chaotic pattern found in other elements. Indeed, it was hard pressed to come up with enough different frequencies of oscillation to explain the complexity of most line spectra. Nonetheless, these patterns were regarded as offering the best clues to the actual structure of the atom. The first model that came up with quantitative predictions would be bound to gain widespread acceptance.

But while the model builders were struggling with the problem of how to get the right light frequencies from the plum-pudding atom, a surprise exper-imental result from the Manchester laboratory of Ernest Rutherford indicated they were betting on the wrong horse. Rutherford was such a towering figure in the physics of his time that it is fitting to pause here and introduce the man.

ERNEST RUTHERFORD

In the 1920s, when he was at the zenith of his fame and recently elevated to the peerage as Lord Rutherford of Nelson, Ernest Rutherford was told by an envi-ous colleague that he was "lucky to be riding the crest of a wave." Rutherford retorted, "Lucky, nothing!—I *made* the wave." While this rebuke hardly shows an excess of modesty, it was perfectly justified. Practically everything known about radioactivity and the atomic nucleus, and by then a great deal was known, had come from the work of Rutherford, his students, and his coworkers.

Rutherford's fierce competitiveness never allowed him to forget the few important discoveries in his chosen field that did not bear his name. Not since Faraday had one researcher so completely dominated an experimental discipline. There was no way to beat Rutherford—you just had to join him. Young physicists flocked to his laboratory, and in an environment where several startling discoveries per year were regarded as commonplace, they stretched and developed their talents to the full.

It wasn't even safe to stay out of his field; if an exciting problem arose in any area of physics, Rutherford might very well pounce on it. For other scientific disciplines he had nothing but disdain: "In science, there is only physics; all the rest is stamp collecting."

Rutherford's fame began almost the moment he stepped off the boat in England in 1895, a raw colonial from the remotest outpost of the British Empire, Christchurch, New Zealand. Some clever studies on the response of magnetic iron to radio waves, conducted under unbelievably primitive conditions in a converted cloakroom at Canterbury College, had earned the ambitious 24-year-old a scholarship to Cambridge as a research student at the Cavendish.

J. J. Thomson permitted a great deal of independence to the junior staff, and in this free environment Rutherford quickly made his mark. In just three years he had attracted enough attention to be offered a chair at McGill University in Montreal, Canada. Though a native-born Englishman might have regarded this as an exile with a dubious future, as a colonial Rutherford had few qualms about taking his chances at a fast-rising institution that was already one of the best in the Empire outside the mother country.

Despite Rutherford's youth, his own reputation and the praise of Thomson were sufficient to attract a stream of first-class assistants from Britain, the United States, and Europe, as well as Canada itself. The productivity of his laboratory over his nine years in Montreal was phenomenal, and in 1907 he returned to England to assume a chair at the University of Manchester. Later he was to cap his career by succeeding Thomson as the director of the Cavendish, but it is on his Manchester days that we will focus, for it was there that he administered the coup de grâce to the plum-pudding atom.

THE CANNONBALL IN THE HAILSTORM

Observing what happened to radiation when it passed through matter had been just part of Rutherford's bag of tricks at McGill, where his primary task had been to identify the composition of the radiation. This problem solved (and the 1908 Nobel Prize earned), Rutherford had the insight to guess that the technique might be turned around. The now well-understood radioactive emanations might serve as a probe to see what was inside the atom. He had noticed that alpha radiation, which he had shown to consist of helium atoms with two electrons stripped away, was deflected somewhat when passing through thin sheets of mica.

This was a small effect that might well have been overlooked. But Rutherford realized that his alpha particles were too heavy and too fast-moving to be budged from their paths except by a strong electrical force. Careful measurements of the deflection could be used to reason back to the size of the force, which might in turn give a clue to how electrons were arranged in an atom.

He had powerful new tools to bring to this task, for Rutherford had made the detection of individual particles of radiation a fine art. On a fluorescent screen such as that on a present-day TV picture tube, an alpha particle made a flash barely visible under a low-power microscope. In Rutherford's own laboratory, his German assistant Hans Geiger was perfecting the electrical counter that bears his name. Thus, while the controversy over the reality of atoms raged, Rutherford had no doubt. When asked over a dinner table whether he believed that alpha particles really existed, Rutherford bellowed: "Not exist— not exist! Why I can see the little beggars there in front of me as plainly as I can see that spoon!"

But like most of his contemporaries, Rutherford had little doubt that Thomson had found the right picture of the atom. A mere corroborative experiment, and one that might prove difficult to interpret, was hardly worth his personal attention. However, there was a new student, Ernest Marsden, looking for a

Rutherford *(right)* and Geiger in the Manchester laboratory.

research topic. A check on whether there might be anything interesting in alpha scattering would make an ideal assignment for Marsden to cut his teeth on.

The prospects for the experiment were interesting but hardly exciting. An alpha particle approaching one of Thomson's plum puddings, as shown in Figure 14-4, would experience no force until it got very close to or inside the atom, for the negative electrons would balance the positive charge. Once inside, the forces would be considerable, but they would be exerted mainly by the electrons. Since these were many times lighter than the alpha particle, the electrons rather than the heavy, swift projectile would be the most disrupted by the encounter.

It would be like a cannonball fired into a hailstorm. After traversing many atoms, the cumulative effect of many small encounters with electrons might have deflected the alpha particle a bit, but no large deflections could be expected. If the deflections resulted from many small scatters, it seemed unlikely that much detailed information about the structure of the atom would be retained. At best, the experiment might give an estimate of how many electrons each atom contained, which was still very much an open question.

The task Marsden faced is illustrated in Figure 14-5. Inside a vacuum chamber (to prevent atoms in the air from interfering), he had to place a thin tube containing a source of alpha radiation. This produced a narrow beam of alphas emerging from the tube. Their target was a thin sheet of gold leaf. Gold was chosen because since medieval times craftsmen had mastered the art of hammering this soft metal to an incredible thinness; good gold foil is translucent. This was essential, because even a sheet of cardboard is sufficient to stop a beam of alphas (it is the more penetrating beta and gamma particles that are primarily responsible for the fearsome reputation of radiation).

FIGURE 14-4. Passage of an alpha particle through a plum-pudding atom.

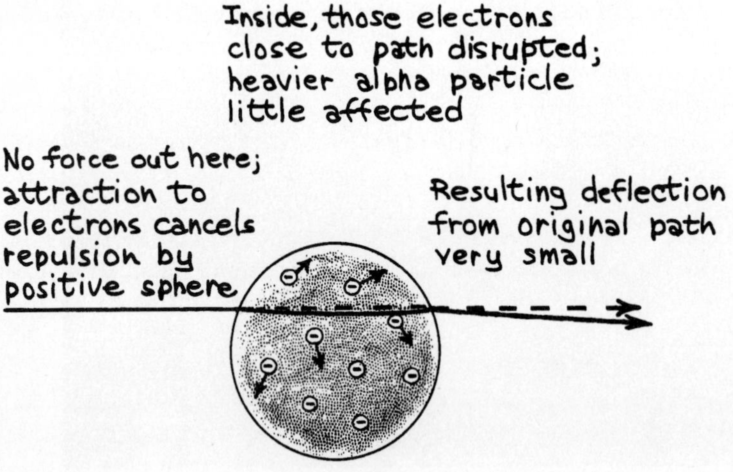

Inside, those electrons close to path disrupted; heavier alpha particle little affected

No force out here; attraction to electrons cancels repulsion by positive sphere.

Resulting deflection from original path very small

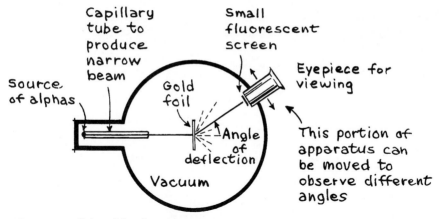

FIGURE 14-5. Geiger-Marsden apparatus.

Marsden had to patiently count the tiny flashes produced when the alpha particles struck a fluorescent screen. To see these flashes reliably required long hours of tedium in a completely dark room.

The first results were not terribly surprising. On the average, the alphas were deflected by only a few degrees. But a very few of them, perhaps one in a thousand, were deflected through substantial angles. Fewer still—about one in ten thousand—even came off backwards!

Again, many physicists might have been content that the average scattering was reasonable. Any number of spurious effects might have produced those one in ten thousand unexpected flashes. But Rutherford had a hunch they were real. He asked Geiger, Marsden's immediate research supervisor, to give the experiment some personal attention.

By 1911, though the data were still crude, Rutherford was sure the results ruled out the plum-pudding atom. Instead, he suggested that the positive charge on the atom might be confined to a tiny region that he called the *nucleus.* The large-angle scatters came from single close encounters with this nucleus.

With this picture of the atom, shown in Figure 14-6, it is easy to explain both the small average scattering angle and the occasional large one. With the atom mostly empty space, the alphas rarely came near a nucleus. Those that did would experience tremendous forces, since Coulomb's law gives a force that varies inversely as the square of the distance. Since the nucleus contains most of the mass of the atom, the cannonball is meeting a bigger cannonball off which it can recoil backward. So the small average deflection comes from the fact that most of the alphas traverse the gold foil without ever getting near a nucleus. The few large deflections come from the rare near misses. More importantly, it was possible to exactly calculate the pattern of this type of scattering. In the Thomson model, the scattering resulted from many tiny

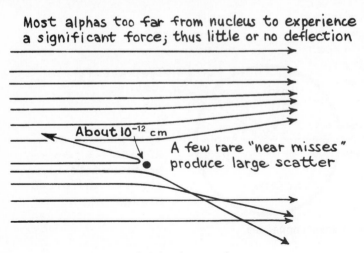

Most alphas too far from nucleus to experience a significant force; thus little or no deflection

About 10⁻¹² cm

A few rare "near misses" produce large scatter

FIGURE 14-6. Scattering of alphas by a nucleus.

deflections from encounters with individual electrons. The result would thus be a pattern that followed the familiar bell-shaped curve of random processes. But in Rutherford's picture the paths of the alphas could be calculated exactly from orbit theory. The result could be expressed in a simple mathematical statement: *the number of alphas per minute hitting the fluorescent screen at some particular angle was inversely proportional to the sine of one-half the angle, raised to the fourth power.*

Though we will encounter this formula several times again, it is not, in itself, terribly significant. What does matter is that it fit the data well enough to rule out the plum-pudding model.

We now know that Rutherford's alphas had in fact nearly touched the nucleus, which has a radius of 8 fm (femtometers, or 10^{-15} meter), about 5000 times smaller than the radius of a gold atom.

Table 14-1 gives the actual data as they appeared in an article in *Philosophical Magazine* (vol. 25, p. 604), a physics journal whose name betrays its origins in the era when physics was still "natural philosophy." It is the third column of this table that spelled the death of the Thomson atom. Though 4000 times as many flashes are seen on the screen at 15 degrees as when it is placed at 150 degrees, dividing the measured numbers of the second column by the computed ones in the first column results in a number that is nearly the same for all measurements. The differences between values in the third column merely reflect the fact that the data are only accurate to about 15 percent.

The values of the numbers represent estimates of the number of flashes per hour seen on a standard screen. In the backward direction, counts extended over many hours, while at small angles the flashes came so quickly that a smaller screen had to be used. What is important is that

TABLE 14-1

Angle of Deflection θ, Degrees	Theoretical Scattering Rate $1/\left(\sin \dfrac{1}{2}\,\theta\right)^4$	Number of Flashes Observed	Col. 3 ÷ Col. 2
150	1.15	33.1	28.8
135	1.38	43.0	31.2
120	1.79	51.9	29.0
105	2.53	69.5	27.5
75	7.25	211	29.1
60	16.0	477	29.8
45	46.6	1,435	30.8
37.5	93.7	3,300	35.3
30	223	7,800	35.0
22.5	690	27,300	39.6
15	3445	132,000	38.4

dividing by the theoretical rate gives numbers that do not vary a great deal. The remaining small variation was of no importance, for with Thomson's atom one could have waited all day without seeing a single flash beyond 75 degrees!

The Geiger-Marsden data serve to illustrate that while physics is often described as a "precise" science, most experiments are no more precise than they need to be. The discrepancy between the predictions of the plum-pudding and nuclear atoms was so enormous that a 15 percent measurement was enough to settle the issue. To strive for greater precision would have been a waste of time and effort.

In a more modern paper, the data would be presented in a graph, as in Figure 14-7, to show the agreement between theory and experiment more vividly. In order to graph such a wide range of values of number of counts, a logarithmic scale is used: i.e., the vertical height of each point on the graph represents the logarithm of the number of counts. This has the effect of compressing the larger numbers, so that the region 1000 to 10,000 occupies no more space on the graph than the region 100 to 1000, rather than taking up ten times as much space. It is a common device to represent graphically numbers that vary over a wide range.

This is the first time in this book we have presented the detailed results of an experiment. It was done in part because of the historical importance of the experiment and also because it came in the era when reporting the data and experimental details, rather than merely the conclusions, became standard practice. But this particular experiment has further historical significance. It was the first demonstration of the power of particle scattering as a tool for studying the forces that operate in the subatomic world. Most of the experiments conducted today with nuclei and subnuclear particles are in a sense variations on a theme by Rutherford, Geiger, and Marsden. Shooting things at

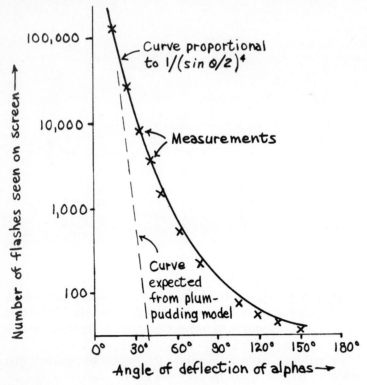

FIGURE 14-7.

atoms and nuclei and studying where they go after the collision is one of the few probes physicists have to get at the workings of matter on the submicroscopic level.

Of course, Rutherford's work raised more questions than it answered. If the positive charge and most of the mass were concentrated in a tiny core of the atom, where were the electrons? Rutherford himself had no idea how to proceed. If the theorists said it couldn't work, let them figure out why it worked nonetheless. Rutherford dealt in experimental fact.

A MOMENTOUS MEETING

There was at least one young theorist who seized upon Rutherford's results with delight. His enthusiasm would eventually make Niels Bohr the guiding spirit in the development of the new physics, the one who dared to make the final break with three centuries of physical thought.

Bohr had come in the autumn of 1911 to the Cavendish Laboratory, to cap his Danish training in theory with an exposure to the British style of experimental physics. He shared a prejudice that a really good physicist ought to have some experience in the laboratory, and there was no place better than the Cavendish to get it. Though he arrived highly recommended and bubbling with enthusiasm, this boyish 26-year-old found Cambridge no bed of roses. One source of annoyance was the petty arrogance of the English academic tradition. His easy-going character rebelled against the stuffiness of an English university town. In a letter home, he complained that his tutor had presented him with "a whole book" on the dos and don'ts of academic protocol.

In the laboratory itself Bohr fared little better. The Cavendish had grown in response to Thomson's reputation and was by then far too large for him to handle. Bohr described the atmosphere of the laboratory as "a state of molecular chaos." Finally, Thomson had shown little interest in Bohr's doctoral thesis, which was based in large measure on Thomson's own work on electrons. Bohr had translated it into English, hoping that his host could help find a publisher that would guarantee a wider audience than was available in Denmark. But it remained on the busy professor's desk, unread.

When Rutherford visited Cambridge for a reunion of Thomson's "old boys," he was introduced to Bohr. By a curious circumstance, the young Dane's name rang a bell. Bohr's brother Harald, who was to become an important mathematician in his own right, was at that time better known as "the shock-haired Dane," star of the Danish football (soccer) team that had unexpectedly won the silver medal at the 1908 Olympic Games in London. Though Niels had never advanced beyond reserve status on his college team, Rutherford, an avid sports fan, referred to him from that day forward as "that football player." An invitation was extended to visit Rutherford's new Manchester laboratory. Bohr returned to Cambridge with stars in his eyes, preaching the gospel of the planetary atom.

To Thomson this was the last straw. It was bad enough that his most illustrious student was challenging his model of the atom. That Rutherford should seduce a guest at the Cavendish into his heresies was too much. By mutual agreement of all parties, Bohr left in April 1912 to finish out his English sojourn at Manchester.

The industrial Midlands were a far cry from the formality of Cambridge. The laboratory was young, the university was young, and above all Bohr was in day-to-day contact with physicists as young and enthusiastic as himself. At the helm was Rutherford, who would stride through the laboratory every morning, giving encouragement to the students and singing "Onward Christian Soldiers" at the top of his voice as he made his rounds. Religion was not his motivation, it was simply the only song he could remember. This was hardly the sort of place where one had to apologize for one's crazier ideas. Bohr was in his element.

This was the beginning of a deep friendship and a long collaboration. Bohr and Rutherford shared a love of vigorous physical activity and a robust sense of humor: otherwise, they had little in common. With a deep mutual respect that grew throughout the years, they above all others shaped the physics of the first half of the twentieth century.

SCIENTIFIC GENEALOGY

A scientist, like the biblical patriarchs of old, can have many offspring. The great ones have their pick of doctoral and postdoctoral students, and can easily train fifty or more in a working lifetime. With the reputation of their mentors behind them, these students have an advantage in finding choice positions, with a chance to make a reputation and attract their own legion of students. And the generation time is short: a professor's early students are only a few years younger than their teacher. Within a few decades, the intellectual descendants of one influential professor can dominate a discipline.

In this fashion, as well as by their ideas and discoveries, Bohr and Rutherford had a telling impact on scientific generations to come. Both drew students from around the world, who became the founders of atomic research in their home countries.

At a recent international conference, about forty physicists went to dine at a country inn in Hungary. They represented the United States, several nations in eastern and western Europe, and East Asia. With them was a historian of science working on the scientific "genealogy" of physicists. As an after-dinner diversion, he worked out the ancestry of all those present. A large majority of the experimenters could be traced back to Rutherford,* and all but one of the theorists was a descendant of Bohr.

Summary

Early speculation on atomic structure favored the so-called plum pudding, electrons embedded in a sphere of positively charged material. Its rival, the "planetary" atom, was unstable and gave no natural explanation of light spectra. But using alpha particles as a probe, Ernest Rutherford discovered that most of an atom's mass is concentrated in a nucleus about ten thousand times smaller than the atom itself. A lucky accident brought young Danish theorist Niels Bohr to Rutherford's laboratory to help resolve the problem. These two figures had a strong influence on later generations of physicists.

*The author of this book, who was present at this gathering, is "third-generation" Rutherford.

CHAPTER 15

The Atom and the Quantum

Hail to Niels Bohr from the worshipful Nations!
You are the Master by whom we are led,
Awed by your cryptic and proud affirmations,
Each of us, driven half out of his head,
  Still remains true to you,
  Wouldn't say boo to you,
Swallows your theories from alpha to zed,
  Even if (drink to him,
  Tankards must clink to him!)
None of us fathoms a word you have said!
—GEORGE GAMOW

Relativity, like the goddess Athena, was born full-grown and fully armed. The quantum theory, in contrast, had a long and troubled childhood. Born in 1900, it took more than ten years to mature to the point where most physicists realized that something new and important was happening, and was nearly thirty when it finally reached adulthood by becoming logically coherent.

Newtonian physics, even as modified by relativity, is above all a science of *continuity* and *determinism*. Nature shows no unexpected fits and starts—all change is smooth, gradual, and the inevitable consequence of definable causes. In the quantum theory, on the other hand, nature on the atomic scale is not only discontinuous but fundamentally unpredictable. It is ironic that this theory entered physics through the study of *incandescence*, a phenomenon that is neither atomic nor discontinuous, and which seems on first inspection to be as orderly as anything in Newton's universe.

MAX PLANCK'S DESPERATE ASSUMPTION

The problem that led to the birth of the quantum theory was the continuous spectrum of light that is produced when a solid or liquid is heated to glowing, as in the filament of a light bulb or a bar of red-hot steel. The phenomenon is not atomic, because the atoms are closely packed and in continuous interaction with one another, completely disrupting their natural way of producing light. When atoms are isolated, as in a gas, each element has its characteristic line spectrum. But in a more condensed state, all materials produce pretty much the same incandescent glow.

181

The spectrum of incandescent light is certainly continuous: light is emitted at all frequencies. The relative brightness of the different colors depends on the temperature. As an object is heated, it first glows red, but as the temperature rises the color shifts toward the blue. This effect became an important research area in the last two decades of the nineteenth century, partly as the result of the success of Edison's incandescent electric light. Experimenters measured the energy radiated at various frequencies, with the results shown in Figure 15-1.

Understanding this spectrum from a theoretical standpoint was viewed as a problem in statistical mechanics, which had already successfully dealt with the conversion of heat into other forms of energy, such as in steam engines and in chemical reactions. The theory had shown that the detailed properties of the materials involved in these transformations were of little importance, as seemed to be the case with incandescent light.

The first attempts showed encouraging results. It proved simple to account for the fact that the total energy radiated increased as the fourth power of the absolute temperature; the theory also explained why, when an object is heated, the color of light emitted changes from a dull red through orange to white and on to blue, as the temperature rises. But the exact spectrum shown in Figure 15-1 eluded mathematical description.

Just before the turn of the century, Max Planck of the University of Berlin took on this problem. Planck was a true professional in his field, the first German physicist to specialize in theory from the start of his training. He also

FIGURE 15-1. Incandescent-light spectrum

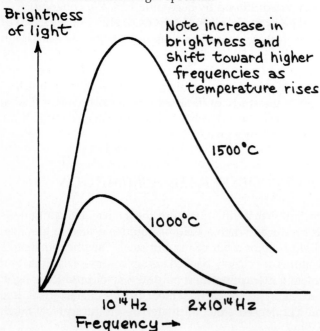

enjoyed the advantage of proximity to the leading experimenters in this area, who were working at his university. Thus he had access to their results before publication, giving him a head start over his competitors. Given this advantage, he decided to let the data guide him to the right formula.

His search was rewarded early in the first year of the new century. But finding an empirical formula that fits the data but is not based on any underlying theory does little to advance our understanding. Statistical mechanics was a well-developed theory, and Planck set high standards for his own work. He felt obliged to come up with a rigorous derivation of his formula.

Moreover, the formula contained an embarrassing adjustable parameter, a kind of "fudge factor" that could be determined only by tinkering with its value until the predictions of the formula fit the measurements. Today this mysterious parameter is known as *Planck's constant*, is designated by the symbol h, and ranks with the gravitational constant G and the speed of light c as one of the fundamental factors that determine the nature of our universe. Planck knew he would either have to derive this number from statistical mechanics, or insert it through some new hypothesis.

By the end of the year, Planck had found his rigorous derivation. But it brought in h through an assumption that looked terribly implausible. He was forced to assume that the conversion of heat into light could not occur in any amount whatsoever. Just as every currency in the world has a smallest coin, the conversion of heat to light had a smallest unit of exchange. The size of this unit depended on the frequency of the light produced, and it came to be called a *quantum*. For light of frequency ν the energy unit was

$$E = h\nu$$

which ranks with $E = mc^2$ as one of the most significant formulas of the twentieth century. The introduction of this formula at a meeting of the German Physical Society on December 14, 1900, is usually taken as the birthdate of the quantum theory.

With this assumption, discontinuity had entered physics. Never before had an important physical quantity been restricted to a discrete set of values. Planck himself was far from sure the quantum should be taken seriously. Perhaps there was some way to eliminate the rule, or explain it away. The quantum might be no more than a way station on the route to a deeper understanding, and many physicists working on the problem expected it to eventually disappear.

But both the formula and the idea of the quantum survive to this day. The cosmic microwave background signal, the evidence for the "big bang" discussed in Chapter 12, exactly follows the Planck spectrum for a body of temperature 2.8 K.

EINSTEIN AND 1905 AGAIN

There are always a few bold thinkers who, confronted with a puzzling new idea, will look not for ways around it but for new ways to build on it. As we have seen, hardly anyone was bolder than Einstein in 1905. He was intrigued by the possibility that Planck might have uncovered a general rule that would operate everywhere in the atomic domain, rather than simply for this one phenomenon. One test of this would be to see if it applied to *light itself* in other situations.

Searching the experimental literature for possible examples, Einstein noticed some peculiarities in the *photoelectric* effect. Today this effect has a wide range of technological applications, from TV cameras to automatic doors and solar energy. But in 1905 it was relatively obscure. It had been discovered by Heinrich Hertz as a by-product of his radio-wave experiments. Hertz had found that when strong light from the blue end of the spectrum shines on a metal carrying a negative electric charge, the charge quickly leaks away. If the charge on the metal is positive, or the light is red, there is no such effect.

Following Thomson's discovery of the electron, Hertz's student Philipp Lenard correctly interpreted this effect. Light must be knocking electrons out of the metal. He showed that electrons were given off at a rate proportional to the brightness of the light. But when he measured the *energies* of the electrons, he got a surprising result. The brightness just didn't seem to matter!

Lenard's apparatus is represented in Figure 15-2. Light shining on the cathode liberates electrons, which are collected at the anode, producing a measurable electric current. By charging the anode negatively, so that it repels electrons, the current can be stopped. From the size of the force required to repel the electrons, their energies can be calculated.

By simple reasoning, brighter light meant stronger electric fields that should have accelerated electrons to higher speeds. But Einstein pointed out that if Planck's quanta were little bundles of light energy, things would be different. Since metals are opaque, light must act only at the surface. An electron absorbs one quantum of energy and immediately breaks free before it can absorb another.

To free an electron from a metal, a fixed "ransom" of energy, called the *work function (W)* must be paid. No electron may have more energy than one quantum, minus the ransom *W*, which leads to a formula for the maximum

FIGURE 15-2. Photoelectric effect

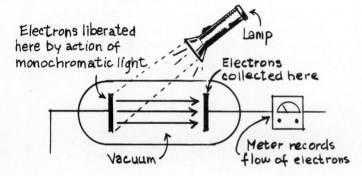

electron energy, $E_{max} = h\nu - W$, illustrated by the graph in Figure 15-3. At low frequencies, the quantum energy is less than W, so there is insufficient energy to liberate an electron. This explains why light from the red end of the spectrum doesn't do the job.

This formula permitted a clear experimental test. The slope of the graph in Figure 15-3 is Planck's constant. If it appears in measurements on the photoelectric effect, which has nothing to do with incandescence, surely Einstein must be right about the broader significance of h. But the experiment required strong light sources that produced pure colors, a difficult achievement in that era. Eleven years were to pass before the American physicist Robert Millikan confirmed the prediction.

This analysis was presented in the first of the five papers Einstein submitted to the *Annalen der Physik* in 1905. The editor who reviewed it was none other than Lenard himself, who was at that time feeling a bit beleaguered. He was considered the world's leading expert on cathode rays, but he had missed the two great discoveries to come out of this work, x-rays and the electron. It may be that Einstein's attention to his photoelectric studies encouraged Lenard to accept the unconventional arguments in this paper and in the two on relativity that Einstein submitted later in that year. In the closing section of this chapter, we shall see Lenard in a more sinister role later in Einstein's life.

The theory of light quanta was the most controversial of all of Einstein's ideas. How could it be reconciled with the overwhelming evidence for the wave theory of light? His quanta must obviously be more like particles than waves, so how could they produce all the wave effects that had been observed in the century since the work of Young? How could light be both a particle and a wave at the same time? Einstein's attempts to struggle with this dual character of light, today a central concept in the quantum theory, reassured nobody.

FIGURE 15-3. Einstein's prediction of the relationship between electron energy and light frequency in the photoelectric effect.

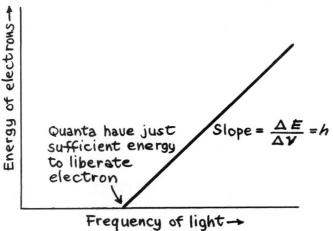

When he was put up for membership in the Prussian Academy of Sciences in 1913, Einstein's sponsors felt compelled to excuse this peculiar lapse of an otherwise obviously brilliant man. Millikan undertook his own photoelectric measurements not to confirm Einstein's prediction but to shoot it down, and nobody was more surprised than he when Einstein turned out to be right. The photoelectric effect, rather than the theory of relativity, was the basis for awarding Einstein the 1921 Nobel Prize, for relativity was still a bit controversial. Nonetheless, not even the Nobel committee fully accepted the light quantum. They were careful to cite the *formula* for the electron energy as the basis for the prize!

In 1923, another American experimenter, Arthur Holly Compton, demonstrated that when x-rays collide with electrons, they rebound with their energy altered exactly as does a particle in an elastic collision. Only then did a majority of physicists accept Einstein's light quantum, and adopt a name that pointed to its particle-like character. Today, a light quantum is known as a *photon*.

This was not Einstein's last word on Planck's quanta. In 1906, he applied the formula $E = h\nu$ to the vibrations of atoms in solids to explain some peculiarities in the absorption of heat at very low temperatures. This came in a period when newly developed refrigerators first permitted research on temperatures close to absolute zero, so Einstein was dealing with a fashionable new topic. It allowed him to show his mastery of statistical mechanics, and to produce results that were of immediate value for the interpretation of experiments, the best way for a theorist to establish his reputation. Though he left the elaboration of this idea to others, he had founded the theoretical branch of a new discipline, solid-state physics. This is now the largest and most practical branch of modern physics.

THE WITCHES' SABBATH

Success in the chemical industry had made Ernest Solvay one of the wealthiest men in Belgium. Still, he was not entirely satisfied with what he had accomplished in life. He would have found it far more worthy to be celebrated as a scientist and philosopher. And so he agreed to sponsor a meeting, in the autumn of 1911, at which the world's greatest luminaries would be assembled to discuss some exciting new ideas. If he could not advance knowledge himself, at least he could lend a helping hand to the process and bask in the radiance of some of the world's most creative thinkers.

This kind of "all-star" meeting was a new idea. The physicist Walther Nernst chose the topic and the guest list. With the reality of atoms now safely established, he felt it was time to alert the world to this hodgepodge of new ideas, the quantum theory.

Lorentz was chosen to chair the proceedings, but Einstein was given the highest honor—the summary talk that would wind up the conference. By then, he was enough of an insider in the academic establishment to be amused by its pretensions.* He was convinced that the assembled dignitaries would simply preen and pontificate for one another's benefit, and would in the end settle nothing. Noting that the conference would take place over Halloween, he referred to it in letters to friends as "this witches' sabbath."

Despite Einstein's misgivings, the first Solvay Conference was an unqualified success, and launched a series of such meetings that continues to this day. It encouraged young theorists to make bold claims for the new theory. One after another, they proclaimed the doom of Newtonian physics at the atomic level. It was time for the small band of quantum physicists to move out boldly and conquer new ground. Einstein's own summary talk, though more restrained, demonstrated impressively how far the new ideas had come.

Though the guest list leaned heavily toward theorists, Rutherford was there. Like many British experimenters he tended to be disdainful of professional theorists, but his own situation now put matters in a different light. He had just made his first public claims for the nuclear atom, knowing that he would have a fight on his hands because the accepted wisdom said it must be unstable. Thus he was receptive to any suggestion that new rules might apply on the atomic level.

So when Bohr arrived in Manchester the following April, Rutherford directed his attention to the printed proceedings of the conference. Though Bohr had heard of the quantum theory, he as yet knew little about it. It was only fitting and proper that he be introduced to it by a document replete with ringing calls to scientific revolution. That sort of adventure was exactly what Bohr was looking for.

BOHR ASKS THE RIGHT QUESTION

Niels Bohr had been raised in the comfortable, supportive bosom of one of the leading families of Denmark's liberal intelligentsia. His father, professor of physiology at Copenhagen University, had used his professorial prerogatives to admit women to his University for the first time. Among the earliest had been Ellen Adler, daughter of a leading Jewish banker who was also a liberal member of the Danish parliament. Christian Bohr was struck by her intelligence and charm, and within a few years they were married.

The Bohr and Adler families were a force in Danish cultural life. Ellen's sister Hannah had studied under the philosopher and educational reformer John Dewey in the United States, and came home to found the first "progressive" school in Denmark. These values held inside the family: the Bohr brothers and their older sister Jenny were encouraged to express themselves freely, to seek their own directions in life, and to participate in organized sports, then considered a "liberal" idea.

---

*In later years, when he settled in the United States, Einstein wrote back to friends in Europe describing Princeton as "a quaint ceremonious village of puny demigods on stilts."

The Bohr home served as the social center for a lively circle of philosophers, scientists, and writers, and the children were encouraged to participate in adult discussions. Coming as they did from a small nation with a difficult language, educated Danes of that era tried to keep one foot in the English-speaking and the other in the German-speaking world, so young Niels was exposed to as broad a range of ideas as could be found anywhere in Europe.

Bohr thrived in this free environment, but always had a great deal of difficulty expressing himself verbally. To start with, he had an exceptionally soft voice, which sometimes made it difficult to hear what he was saying. And he tended to speak in half-constructed poetic images, sharing his stream of consciousness with his baffled listeners. In later years, Bohr explained this trait as essential to his manner of thought: "Never express yourself more clearly than you think!"

Nonetheless, Bohr's own father saw something through the barrier of muddled words, and pronounced a judgment that was borne out in time: Harald was the quick one, and would go far. But Niels was deep, and would do great things.

The Danes preserve the charming custom of holding oral examinations for the doctorate as public events. Niels and Harald defended their theses before packed houses, to rave reviews in the daily press! With such a background, Niels Bohr never hesitated to believe he could turn the world upside down with the products of his mind.

Rutherford was bound to be flattered that such a well-recommended young man viewed his nuclear atom with enthusiasm, rather than the dismay expressed by most of his own contemporaries. Whatever may be the value of scientific journals, emotional commitment to ideas is best transmitted by personal contact. Elsewhere in the world, Rutherford's nuclear atom seemed a possibility. At Manchester, Bohr was in the midst of a group of talented young researchers who took it as established fact. The planetary atom had to be made to work, and the quantum theory just might hold the key. In his four months at Manchester, he had three key insights that set him on the right path.

The first was the key to the stability of atoms. If the essence of the quantum theory is that a system is allowed only certain values of energy, there could be a *lowest* energy level, the smallest possible orbit, below which the electron could not descend. There would be other allowed energies, represented schematically in Figure 15-4. The lowest energy is the *ground state*, in which an undisturbed atom is normally found. The higher levels are *excited states*, which appear only if the atom is disturbed by a process that can insert energy, such as a violent collision. An atom would not stay very long in an excited state. It would return to the ground state in a series of steps. Both the ground state and the excited states are bound states, so the total energy is negative. An electron with positive energy is free to wander away.

Bohr then moved to his second insight. Something must set the size of the first orbit, and thus the size of the atom itself. Bohr reasoned that it must depend on m and e, the mass and charge of the electron. But if the quantum theory was to play a role, Planck's constant must also enter the picture. He

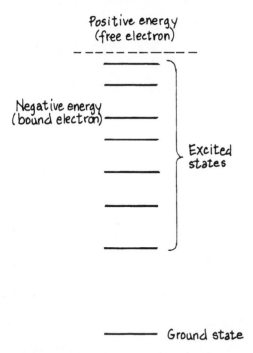

FIGURE 15-4. An energy level diagram.

then noted that the combination h^2/me^2 is a *length* unit roughly in the range of known atomic diameters.

Others had already noticed that combination, but had understood it differently. They had tried to explain away Planck's constant as a consequence of the size of the atom and the charge and mass of the electron. Bohr had reversed their line of reasoning. It is not the job of the atom to explain the size of Planck's constant, but for the constant to explain the size of the atom!

Bohr's third insight was that an atom with a restricted set of energy levels could change its internal energy only by moving from one allowed level to another. This would be the way it could emit light. Here was a natural explanation for line spectra, for the relation $E = h\nu$ could then be stood on its head. Up to then, the formula had set the size of an energy quantum when the frequency was known. But in Bohr's picture, the energy levels would be set by some other rule. The frequency was determined by dividing the change in energy by Planck's constant.

The task ahead of him was clear: find a rule that gave the energy levels. But Bohr had run out of time. He had to return to Copenhagen, for his wedding day had been set. But he did call off the planned honeymoon in Norway, bringing his bride back to Manchester. Starting his first teaching job in the fall, he had little time to concentrate on the problem. He played with orbit calculations, but had no idea which way to go. It had not occurred to him that atomic spectra might provide a clue.

Bohr's engagement photo, taken shortly before his departure for England.
(Courtesy of American Institute of Physics.)

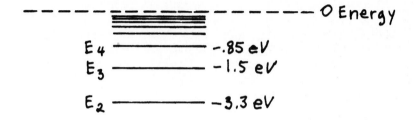

$$E_4 \quad\text{------} \quad -.85\ eV$$
$$E_3 \quad\text{------} \quad -1.5\ eV$$
$$E_2 \quad\text{------} \quad -3.3\ eV$$

$$E_1 \quad\text{------} \quad -13.2\ eV$$

FIGURE 15-5. Hydrogen energy levels.

Early in February 1913 H. M. Hansen, an old classmate of Bohr's and a spectroscopist, paid a visit to Copenhagen. He reminded Bohr of the Balmer formula, which both had learned in a class long before but Bohr had forgotten. Going back to his notes he found it and, as he later put it, "everything was immediately clear."

It was impossible to miss the parallel between Balmer's formula

$$v = v_0 \left(\frac{1}{n^2} - \frac{1}{m^2} \right)$$

and Bohr's prescription for calculating the frequency of light emitted in a transition between energy levels n and m:

$$v = \frac{1}{h} (E_n - E_m)$$

These formulas look very similar. We can make them exactly the same by setting $v_0 = E_1/h$, where E_1 is the energy of the ground state, and letting the energy of the nth state be E_1/n^2. The pattern of energy levels suggested by the Balmer formula is illustrated in Figure 15-5.

From the foregoing discussion, it would be natural to assume that Bohr eagerly accepted Einstein's light quantum. Curiously, this was one aspect of the quantum theory that Bohr steadfastly resisted, clinging to this opinion for two years after the Compton experiment had made it untenable.

Bohr correctly guessed that the hydrogen atom contained only one electron, which simplified his calculations. Now his task was to find a new quantum rule, something different from $E = hv$ that led to the right numerical values for the energy levels. Bohr quickly found it, but at this point it is important to break off the narrative and issue a warning.

THE WRONG PICTURE, THE RIGHT NUMBERS

Up to now, all of Bohr's ideas were on the right track, and they remain valid today. In this section we present his *model* of the atom, Bohr's picture of how the electron moves. It cannot be emphasized too strongly that *this picture is entirely wrong, even though it leads to the correct energy levels!* Thus we will touch on it briefly, and then move on to a more satisfactory picture that emerged some thirteen years later.

There is a methodological lesson in this that should be taken to heart. Model building is a dangerous business, particularly when one is guided by very good experimental data. If a model can be devised that explains the data, it seems only reasonable that that model, or something very much like it, must be the way things really are. But this may prove a snare—another entirely different model may serve just as well.

With only a single electron to consider, Bohr decided to adopt Newton's derivation of Kepler's laws as the basis for his model. He then added three assumptions, *all three* of which were later abandoned:

1. The energy levels are those *circular orbits* in which the electron's angular momentum is an integer multiple of Planck's constant divided by 2π, i.e., $mvr = nh/2\pi$.
2. Maxwell's theory does not apply to the electron's motion, so it will not radiate light while orbiting.
3. The electron moves from one orbit to another by an instantaneous process called a *quantum jump.*

The combination $h/2\pi$ appears so often in the quantum theory that to simplify formulas it has been assigned a symbol of its own, $\hbar$, which is pronounced "aitch-bar." Though Bohr's orbit rule did not endure, his surmise that $\hbar$ is a natural unit for angular momentum was correct.

To give Bohr full credit, he was the first to acknowledge that he was on thin ice. The old physics had three centuries of work to back it up; all he had was Rutherford's experiments and the Balmer formula. So he warned his readers not to take the model too literally. It was a statement in the language of the old physics. Its meaning would be clear only when it was restated in the language of the new physics that was bound to emerge, and which he correctly guessed would have little in common with Newton's system.

It took a special kind of mind to accept the contradictions and confusions of this kind of reasoning and still forge through to a conclusion, knowing that many mistakes may have been made along the way. Einstein, with his passion for consistency and logical clarity, could never have done it. He was astonished and delighted by Bohr's "unique instinct," and praised his achievements as "the highest form of musicality in the sphere of thought."

NUMBERS ARE POWERFUL CONVINCERS

Nobody could dispute the fact that Bohr's theory came up with those marvelous little convincers, experimental numbers. Frequencies of spectrum lines can be measured to great accuracy, and Bohr hit them right on the head. Let us summarize his quantitative results. If complicated mathematical expressions make your eyes glaze over, read the text and ignore the formula, which is there primarily for illustration.

The radius of Bohr's ground state orbit is in fact the length he originally guessed at, with h replaced by $\hbar$, $\hbar^2/me^2$, which is around 5×10^{-11} meter. This leads to the following expression for the energy of the nth state:

$$E_n = - \frac{me^4}{2\hbar^2 n^2} = - \frac{13.6}{n^2} \text{ eV}$$

Here we have introduced the *electron-volt (eV)*, the unit of energy commonly employed on the atomic scale. It is equal to 1.6×10^{-19} joule. In these units, $h = 4.14 \times 10^{-15}$ eV-second and $\hbar = 6.6 \times 10^{-16}$ eV-second.

The point of displaying this curious formula is that it is, in fact, a peculiar-looking combination of tiny quantities. That it turns out to give the right answer could hardly be a mere coincidence. Any model that replaced Bohr's would have to duplicate this achievement, and leave this formula unchallenged.

THE BATTLE IS JOINED

Bohr's theory was greeted with considerable skepticism. Rutherford, to whom Bohr sent the paper for criticism before submitting it for publication, was not yet prepared to stake *that* big a change in the foundations of physics on the slender base of his alpha scattering experiment. But Bohr seemed determined to go public with the theory, so Rutherford acquiesced. A number of readers of the article expressed bafflement that Planck's quantum, a creature of statistical mechanics, could be relevant to spectroscopy, which dealt with individual atoms. As Bohr himself realized, he had taken only a first step on the road to the new physics, and more work had to be done.

But suddenly it was August 1914, and there was nobody left to do the work. With the assassination of an Austrian archduke in the little-known Bosnian city of Sarajevo, young scientists who had been working side by side suddenly found themselves conscripted into rival armies, to face one another across fields of barbed wire.

Rutherford's Manchester laboratory was quickly denuded of the students and assistants who had made it hum. Rutherford patriotically resigned himself to research on submarine detection, while his friends and former coworkers in the German laboratories worked to frustrate his efforts. Rutherford had good personal reasons to reflect bitterly on the futility of the slaughter. His own prized student and protégé, Henry Moseley, died in the disastrous British attack on the Dardanelles. Moseley's greatest achievement, just before he left for the army, had been to show by experiment that the Bohr theory explained the production of x-rays. But this did not exempt him from the horror that cut down a major share of European youth of his age.

PHYSICS IN THE TWENTIES

When the surviving young physicists returned from the trenches at the end of 1918, the intellectual world was in no mood for caution. The three great emperors who had dominated European politics had lost their thrones. European civilization had itself been discredited in the eyes of many intellectuals. The Bolshevik revolution in Russia had aroused the passions of the whole world.

Physics often seems the most insulated of intellectual endeavors. Perhaps it was mere coincidence that classical physics seemed on the verge of ultimate triumph in the complacent 1890s, only to be replaced by a new physics born in the turbulent 1920s. In purely scientific terms, the time was ripe. Not all sciences experienced a similar flowering. And great science is usually done by people who, for the time being at least, are thinking of little else.

Yet one can imagine the mood of a young physicist whose diversions might include the plays of Bertolt Brecht, the music of Paul Hindemith, the novels of Thomas Mann, and Dada art exhibits. The European educated elite was still small, and a scientist was bound to have neighbors who were surrealist painters, radical poets, Bauhaus architects, or devotees of the daring psychological theories of Sigmund Freud. The mood of the cabarets of Berlin and Munich, a blend of high spirits and decadence, quickened the pulse. Was this the time to dwell contentedly on the ancient traditions of physics, to add one's little bit to the great edifice built on the foundations of Newton?

Science had ridden high in imperial Germany, a state that claimed to rest its legitimacy on scientific principle. In the new Weimar Republic, science lost much of its status. To keep afloat, the great research institutes of the German universities had to draw a major share of their students, and even some of their financial support, from abroad. It should be noted that a similar situation exists in science education in the United States today.

Physics was in particular disfavor because of the role assigned it in one of the most influential books of the era, Oswald Spengler's *Decline of the West.* Spengler condemned Western civilization for having made a Faustian bargain to gain mastery over nature. The dead hand of determinism, epitomized by the laws of Newtonian physics, had traced the West's signature in blood. Echoing the romanticism of a century earlier, Spengler proclaimed mysticism, intuition, and chance the stuff of life on which a new civilization would arise.

The quantum physicists did not take this criticism lying down. There was mystery aplenty in quantum jumps and intuition in their methods. A new physics was aborning, and determinism was finished.

Einstein and Planck were wary of the new mood. By all means build the new physics, they warned, but to capitulate to the forces of unreason was to let in the hounds of hell. Already these were very much in evidence on the streets, with their brown shirts and swastika armbands, and listed among their targets was the "Jewish science" of theoretical physics. Philipp Lenard, one of the few eminent scientists to join the Nazi party, organized pickets and hecklers to disrupt Einstein's public appearances, and called for the creation of an "Aryan physics" in which experimenters would rule the roost and theories formulated by Jews would be rejected.

Most German intellectuals deplored these excesses and endorsed Germany's first experiment with democracy. But they maintained a haughty, aristocratic disdain for the give-and-take of parliamentary politics. The Weimar Republic foundered not so much on the hatred of its enemies as the indifference of its supposed friends.

But in this inhospitable climate, German physics rose to its greatest heights. In the eye of the hurricane sat the great Georgia Augusta University at Göttingen, a provincial university city reminiscent of an American college town. Göttingen had been a center of intellectual ferment and political rebellion since its foundation in the eighteenth century by George II, Elector of Hanover and King of Great Britain.

The key figure at Göttingen was David Hilbert, perhaps the most influential mathematician of the century. He cultivated an atmosphere of intense scientific debate that cut across traditional disciplinary boundaries. A visiting speaker approaching a Göttingen lecture faced an audience that would accept nothing short of daring original ideas. There were discussions long into the night on the shape of the new physics. The students were in no mood to be polite to their elders.

But the mecca of this new religion was unquestionably Copenhagen, and its unchallenged prophet was Niels Bohr. Barely into his thirties, Bohr headed a new institute supported in part by the profits of the venerable and world-renowned Carlsberg brewery. Since Denmark was a small country, it was understood from the outset that this would be an *international* institute, attracting the best young scholars from throughout the world. Bohr was to make Copenhagen nearly as famous for physics as for good beer and rollicking high times. Rumors of an important new development anywhere in Europe brought a reflex reaction to the young guard of physics—catch the next train to the charming and fun-loving city on the Øresund. Only in "The Presence" could the true significance of a new idea be evaluated, and the debates became legendary.

Bohr's work habits enhanced his influence. Unlike Einstein, who craved solitude, Bohr thought best out loud, in a madcap environment full of people to bounce ideas off. Work at his institute might be conducted across a ping-pong table (Bohr was almost unbeatable) or on a tour through the Tivoli amusement park. Diversions included Western movies and the in-house *Journal of Jocular Physics*, edited by George Gamow, in which the latest ideas were promulgated through the medium of Mickey Mouse cartoons. When a visitor remarked that *disrespect* seemed to be the hallmark of his institute, Bohr impishly replied: "Yes, and we don't even take disrespect seriously!"

Bohr's own ideas often began as metaphors, poetic images with meanings his listeners could scarcely fathom, but which encouraged them to let their own imaginations run free. In Copenhagen, the new physics was not so much built as slapped together in a riotous spree of individual and collective creative effort. When the time came to put it all down in solid, rigorous mathematics, there was Harald's own institute, right next door.

The settings were familiar—the institutes of the German and French universities, the Cavendish Laboratory, a bit less stuffy after Rutherford succeeded J. J. Thomson in 1919, the gentle Scandinavian frivolity of Copenhagen. But the mood was new, and every bit as romantic as Hemingway's Paris or Brecht's Berlin.

In Munich, in the same era that witnessed Hitler's beer hall putsch, the waiters at one cafe near theorist Arnold Sommerfeld's institute had peculiar instructions. When the young physicists who passed their evenings there left the marble-topped tables covered with equations, they were under no circumstances to be cleaned; a number of the key ideas of the new physics went straight from those tables to the pages of the leading journal of the quantum-mechanical revolution, *Zeitschrift für Physik*.

HEISENBERG AND PAULI

The story of Sommerfeld's two most illustrious students puts in bold relief the cultural conflicts that wracked Weimar Germany. Werner Heisenberg and Wolfgang Pauli were together in Munich in the early 1920s, and for a decade or so remained close friends, despite striking differences in background and personality.

Heisenberg was physically almost a caricature of the archetypical German—tall, fair, and athletic. He came from a respectable Bavarian family, with a father who was the principal of a gymnasium. Young Werner was taught to value discipline and self-denial. As a student, he was respectful toward his teachers and meticulous in his study habits.

Pauli, by contrast, was a coddled child prodigy from the Jewish community of Vienna, though he was baptized as a Christian. Short and stocky, he

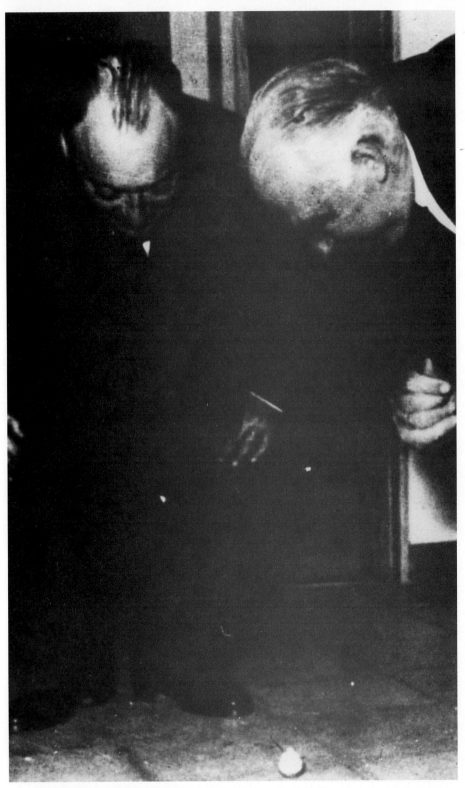

Wolfgang Pauli (left) and Niels Bohr contemplate a spinning top.
(Courtesy of American Institute of Physics)

Niels Bohr and Werner Heisenberg rest during a mountain hike.
(Courtesy of American Institute of Physics.)

was given to self-indulgence, with a weakness for fine wines, imported cigars, and ice cream. He also had a well-deserved reputation for arrogance. If a professor had not made his point clear enough for Wolfgang to understand, he was likely to be ordered to "stop talking nonsense."

Heisenberg was a member of the Youth Movement *(Jügendbewegung)*, a peculiarly German institution that strove for a romantic "renewal in body and spirit" along the lines championed by Spengler. Civilization and rationalism were shackles on the German soul, and young people ought to flee from cities and books and head for the lakes and mountains to commune with the old pagan gods. Theoretical physics was, to say the least, not looked upon with favor.

Pauli's preferred recreations were ones that Youth Movement stalwarts would brand as "degenerate"—late evenings in a smoky cafe or cabaret, with worldly conversations lubricated by generous libations of beer and wine.

As a member of the movement, Heisenberg would spend a weekend camping with his companions beside a mountain lake, rising well before dawn on Monday to board the milk train back to Munich in time to catch Sommerfeld's lecture. Often as not, Pauli would be home asleep, having spent the evening in some low dive and the wee hours working on deep physics problems that went far beyond what the professor was teaching.

Heisenberg set himself the task of "reforming" Pauli, once persuading him to take an extended bicycle trip through the foothills of the Bavarian Alps, sleeping in haylofts or wherever they could. Pauli's reaction was to marvel that an apparently intelligent person would willingly subject himself to such acute discomfort. When Heisenberg offered to reciprocate by accompanying Pauli on one of his nightly prowls, he was rebuffed with "That wouldn't do at all—it's not your kind of life."

Nonetheless, Heisenberg and Pauli maintained a high level of respect for one another and even managed to collaborate on some significant projects, until the rise of Nazism separated them. Pauli saw the handwriting on the wall, and fled Germany for the safety of a post in Zurich well before Hitler's takeover. Heisenberg, always the good German, deplored the Nazi excesses but nonetheless remained at home, faithfully serving the regime. In their rare encounters after the end of World War II, Pauli never felt comfortable with Heisenberg again.

Summary

The quantum theory originated in 1900 through Planck's theory of incandescent light emission, which contained the assumption that energy could not be transformed from heat to light except in discrete units that depended on frequency of the light. Einstein reinterpreted this as a general rule and constructed a theory of the photoelectric effect, the least accepted of his early ideas, but later the basis for his Nobel Prize. Bohr learned of the theory through a conference Rutherford had attended and developed a theory that explained the light spectrum of hydrogen. Its key ideas were that atoms have a limited repertoire of energy states, and radiate light only when changing states. The specific model for these states as circular orbits did not endure.

CHAPTER 16

Particles and Waves

We are trapped by language to such a degree that every attempt to formulate insight is a play on words.

—NIELS BOHR

A research scientist is sometimes compared to a hunter stalking an elusive beast. The physicists who built quantum theory in the early 1920s more nearly resembled a rowdy band of schoolchildren chasing a rabbit across a rocky meadow. The order of the day was "anything goes," and the approach was scandalously ad hoc. Invent a quantum rule, derive a formula, check against the data, and go on to the next problem. An understanding of what sort of physical reality might underlie a successful computation would come in its own good time. The hope was that after enough lucky guesses a pattern might emerge to guide physicists to a deeper level. In the meantime, the new game was just plain fun.

THE BOHR ATOM IS NOT ENOUGH

The central problem in the early 1920s was to make Bohr's model work for heavier elements, which would have many electrons orbiting the nucleus. This was an extraordinary challenge, for even Newtonian physics had been unable to deal with more than two objects at a time, except in cases like the Solar System in which one center of force dominated all the others. In an atom, the forces that electrons exert on each other are comparable to their attraction to the nucleus. To this very day, with the aid of the world's most powerful computers, only approximate calculations of the energy levels of multielectron atoms can be performed.

A new quantum rule developed by Arnold Sommerfeld gave some of the circular orbits a few elliptical companions, each as long as the diameter of the circle. Such an orbit has the same energy as the circular one. Where would all this arbitrary rulemaking end? It seemed that every new problem brought into being a new rule. It was great fun, but was it science? What lay behind all these strange, arbitrary rules?

As is often the case when a group of scientists are deeply immersed in a problem, the key to the muddle came from outside, from a more isolated thinker with the leisure to contemplate the problem in a detached manner. And it came not from Germany or even Denmark but from France, where a less frenzied and more reflective style dominated theoretical physics. The protagonist, Prince Louis Victor de Broglie, was one of the most improbable characters in a drama full of improbable people.

A PRINCE HAS A CRAZY IDEA

Few families in Europe outrank the de Broglies in the *Almanach de Gotha,* the quasi-official register of nobility. The eminence of his lineage is attested to by the fact that Louis bore the title of *Prince* as a mere cadet honor; following the death of his older brother he assumed the higher title of *Duc.* The de Broglies have provided France with diplomats, cabinet ministers, and generals for centuries. One of Louis de Broglie's ancestors even fought on the American side in the War of Independence. Accordingly, de Broglie received the standard humanist education that is the traditional preparation for a role in the French ruling elite, taking a *Licence,* the equivalent of a master's degree, in political science.

But the family also had a modest scientific tradition. Louis' elder brother Maurice was an experimental physicist with a strong enough reputation to be elected to the presidency of the French physical society. He studied x-rays in a private laboratory funded out of his own pocket, one of the last professional scientists to work in this fashion.

Through his brother's influence the young prince took an interest in the work of Einstein. He was particularly intrigued by the possibility of finding a connection between Einstein's two most original ideas. He hoped that relativity itself might shed some light on the problem of the particle characteristics of light implied by Einstein's treatment of the photoelectric effect. Work in this field naturally went slowly for this rank amateur, who had in effect to start his education all over again. He completed his physics *Licence* shortly before the outbreak of World War I.

Returning from five years of military service, Louis decided to see his interest in physics through to the doctorate, with a thesis on the wave-particle problem. But in the meantime, the success of the Bohr model had changed the whole picture in the quantum theory. No longer was the quantization of light the central mystery. The strange limits on the motion of electrons in atomic orbits were even more disturbing. Convinced that the wave-particle duality was the key to the earlier quantum theory, de Broglie wondered whether it might also explain Bohr's restricted orbits. If light could be a particle, why couldn't an electron be a wave?

The idea had many inviting aspects. While it was difficult to imagine a way to force a particle to avoid all but a limited repertoire of motions, similar restrictions on waves were quite normal. After all, musical instruments work because the standing-wave motions allowed on a taut string or in an enclosed column of air are severely restricted. If the electron could in some sense be regarded as a wave, perhaps the Bohr orbits would prove to be standing

De Broglie.
(Courtesy of American Institute of Physics.)

waves and the ground and excited states would become the atom's "fundamental" and "overtones."

But treating an electron as a wave was not quite as simple as treating light
as a particle. Light always travels at the same velocity, so its frequency determines its wavelength. An electron, capable of moving at any velocity whatsoever, would need separate rules for the wavelength and the frequency. $E = h\nu$
might be enough for the photon, but a separate rule giving the wavelength
was essential for the electron.

Reasoning by analogy rather than by rigorous mathematics, de Broglie
decided that if the frequency of a wave is related to the *energy* of a particle, the
wavelength might be related to the *momentum.* Accordingly, de Broglie set
forth, in his celebrated 1924 doctoral thesis, the following formula:

$$p = \frac{h}{\lambda}$$

relating the momentum of an electron to the wavelength λ of a wave associated in some mysterious way with the electron.

Now de Broglie could test his surmise about Bohr's orbits. If they were
really standing electron waves, the Bohr orbits must be the ones in which a
whole number of wavelengths fit, as shown in Figure 16-1. The electron wave
would travel around the orbit, reinforcing itself constructively at each turn,
just as the wave on a music string is reinforced by its successive reflections.
The waves must fit evenly into the circumference of the orbits. Stated mathematically, the allowed standing waves are those for which an integral number
n of waves fit in the circumference of the circle:

$$\frac{2\pi r}{\lambda} = n$$

FIGURE 16-1. De Broglie's picture of a
Bohr orbit.

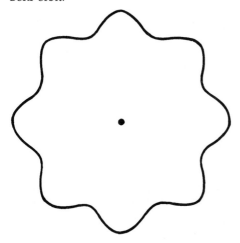

Since his wavelength rule gives $\lambda = h/mv$, we can substitute this value for λ and get

$$2\pi rmv = nh$$

which is Bohr's orbit rule!

To summarize the reasoning, if an electron is somehow associated with a wave, with its wavelength determined by the momentum, then the allowed standing-wave patterns for an electron circling a nucleus are the very same circles as Bohr's allowed orbits.

De Broglie's dissertation was a hot potato for the faculty of sciences of the University of Paris. A thesis, while expected to be original, isn't often *that* original. A solid contribution to an established topic is safer, even for the most brilliant student. And here was a convert from the humanities, writing on a theory not yet well-known in France, promulgating an outrageous idea. De Broglie couldn't explain how the electron, which showed every sign of being a particle, could at the same time be a wave. In his own words, de Broglie characterized his theory as "a formal scheme whose physical content is not yet determined." And apart from the striking coincidence of duplicating the Bohr orbits, no theoretical or experimental justification was offered.

The problem was resolved by de Broglie's sponsor, Paul Langevin, the only member of the committee who was actively working in quantum physics. To support his own judgment he went for an outside opinion, from no less a personage than the great Einstein. The response was encouraging: "It may look crazy, but it really is sound!"

Now matters were working to de Broglie's advantage. In Germany, home of the quantum theory, Einstein himself was promoting the idea of the electron wave. The story has the sort of ending one might find in a fairy tale: the student prince became the first physicist to receive a Nobel Prize for his doctoral thesis.

HELP FROM AMERICA

The money for Michelson's first interferometer was not Alexander Graham Bell's only gift to fundamental science in America. He also founded the Bell Telephone Laboratory, an institution that has earned no less than four Nobel Prizes in physics. Two of these were for the sort of practical accomplishments one might expect from a leading industrial laboratory—the development of the *transistor* and the *maser* (a microwave amplifier that is the ancestor of the *laser*). But the other two fall in the domain of pure science, even though both were by-products of more practical research.

One was the discovery of the cosmic 2.8 K background radiation, which came out of a search in the 1960s for radio noise that might interfere with satellite communications. The other, and the first of the four prizes, came for showing that the French prince was not crazy after all.

Clinton Davisson had been hired by Bell Labs to conduct studies of collisions of electrons with metal surfaces in a vacuum. At that time the *vacuum tube* was rapidly becoming the mainstay of the communications industry, but the science underlying the device was still full of unknowns, so improvements had to be made by trial and error. In a vacuum tube, electric fields control the flow of electrons from one metal electrode to another. Some electrons bounce off the target electrode, creating a cloud of negative charge that disrupts and limits the flow of current.

Davisson found that electrons were reflected much more strongly at some angles than at others. To make matters worse, the preferred angles depended on how fast the electrons were moving. With no theoretical framework to account for this curious result, he simply reported it. But at Göttingen James Franck and Walter Elsasser, who were also working on electron scattering, had seen a copy of de Broglie's thesis. To their eyes, Davisson's results looked like a clear case of wave interference.

Their interpretation of Davisson's result is illustrated in Figure 16-2. If the electron is a wave, only at certain angles would the portions of the wave scattered from different atoms interfere constructively. The most suggestive clue was the way in which the angles for strong reflection depended on velocity. This must result from the changing wavelength of the electron. The spacing of atoms in nickel, the metal used by Davisson, was known. Franck and Elsasser could calculate the wavelengths from Davisson's angles, and also by using the electron velocities and de Broglie's formula. Within the somewhat limited accuracy of the measurements, both estimates gave the same wavelength.

FIGURE 16-2. Wave effects in the scattering of electrons off metals

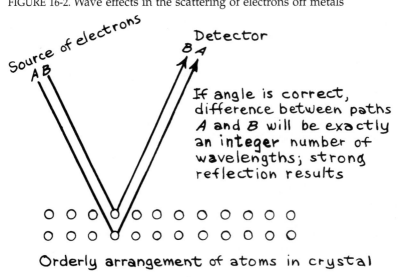

Once again, the value of publishing raw experimental data had been demonstrated. Davisson was hardly likely to have seen a speculative doctoral thesis printed in French and circulated informally. But the German physicists were on the quantum "grapevine" and carried the day for de Broglie's crazy idea.

Davisson's superiors didn't need to ask what Alexander Graham Bell himself would have done in this situation. He was told to forget about the damned vacuum tubes—this could be something big! For two years Davisson refined his work. A trip to a scientific meeting in England brought him into the world of quantum physics. G. P. Thomson, son of the illustrious J. J., was confirming Davisson's results, and both experiments were soon too conclusive to leave any room for doubt. Davisson and Thomson joined de Broglie in the ranks of Nobel laureates, prompting one wag to remark that if old J. J. had gotten the prize for proving the electron was a particle, why shouldn't his son get it for proving it was a wave?

But no measurement could shed any light on just how one object could be both a particle and a wave. This became the central issue in the development of the quantum theory. But while the experiments were nearing completion there were developments on the theoretical front that promised to make the quantum theory, for the first time, a reputable successor to the Newtonian system.

A WAVE EQUATION

Einstein spread the word about de Broglie's thesis to a number of his friends. One of these was Pieter Debye, his successor at the Polytechnic in Zurich, who asked his colleague at the University of Zurich, Erwin Schrödinger, to report on it for their joint seminar. The winds of fate had carried de Broglie's seed to fertile ground. Debye remarked that de Broglie had visualized a one-dimensional wave, but an atom is three-dimensional. The remark did not fall on deaf ears.

Schrödinger was a true son of the city of Vienna, raised in the brilliance and decadence of its last years as the capital of a great empire. Endowed with considerable personal charm and a flair for self-advancement, he had rapidly ascended the academic career ladder by the time-honored route of job-hopping. His reputation was based not so much on his own ideas as on his mathematical virtuosity. He specialized in elegant refinements of the half-formulated works of others, exactly the treatment that de Broglie's idea called for.

At the time he set eyes on de Broglie's thesis, Schrödinger was nearing 40 years of age and was beginning to fear that his colleagues no longer took him seriously. It was not merely his lack of originality that harmed his reputation. He had written on a wide range of topics without ever becoming an expert in any one. His outside interests were equally broad, encompassing poetry, the theater, and Indian Vedantic philosophy, as well as physics and mathematics. Furthermore, he had a cosmopolitan disregard for conventional morality and a penchant for amorous liaisons that proved a drain on his time and energy. Like many European (and especially Viennese) intellectuals of his era, he combined a deep pessimism about the world with a sensual indulgence in whatever poor pleasures it might have to offer. All in all, he seemed spread a bit too thin to have much impact in his chosen profession. Einstein summarized Schrödinger's character in three words: "a clever rogue."

De Broglie's electron wave gave him a new lease on life. Characteristically, he had come close to the same idea four years earlier, only to flit away to new interests. Now he summoned his highest powers of concentration. Holed up in a Swiss ski resort for the Christmas break of 1925, he found the answer in a matter of weeks. There followed six months of astonishing creativity. In four lucid and elegant papers he outlined the new paradigm of *wave mechanics*, a system nearly as all-encompassing and logically self-contained as Einstein's relativity or Newton's mechanics.

What he constructed was a single definitive equation whose solutions would describe the de Broglie wave in any situation in which the force was known. It is reproduced below simply for the historical record; its mysterious symbols make sense only to those with at least two years of training in the calculus. What is important is that it takes the form of a *partial differential equation*. Since the late nineteenth century, physicists had been trained to formulate all theories encompassing three dimensions, from field theories like Maxwell's to Newton's laws themselves, in this mathematical form.

$$\left(\frac{-\hbar^2}{2m} \Delta^2 + V \right) \psi = i\hbar \, \frac{\partial \psi}{\partial t}$$

The symbol ψ is a mathematical representation of the de Broglie wave. In order to satisfy the Schrödinger equation, it must meet the following requirements:

1. The wave must obey $p = h/\lambda$ and $E = h\nu$.
2. The force present is taken into account by means of the potential energy V it produces.
3. Energy and momentum are conserved, in their nonrelativistic form.

A free particle is the simplest case, and one example is illustrated in Figure 16-3. An electron passes through a thin metal foil. When the electron reaches the surface of the metal, it is attracted to the atoms on the surface and it speeds up. Its kinetic energy and momentum increase, so its wavelength becomes shorter. Once inside the metal, the electron is equally attracted in all directions, since it is surrounded by atoms. At its exit, it once again is attracted by surface atoms, slowing it down to its original speed.

The prescription for finding the standing waves allowed in a bound system is more involved, but is quite straightforward. Nowadays it can be turned into an *algorithm* that allows a computer to do the work. Start by making a guess at the particle's energy. At some position inside the atom, subtract the potential energy to get the kinetic energy. From the kinetic energy one can calculate the momentum and obtain the wavelength. This is repeated at many

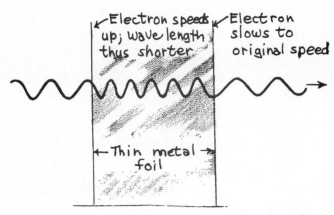

FIGURE 16-3. Wave picture of an electron passing through thin metal foil.

locations, to obtain little pieces of the wave. If these pieces fit together smoothly, the guess at the energy was right and is one of the allowed energy levels of the system.

But the 1920s were well before the computer era, and Schrödinger had to rely on pencil-and-paper mathematics. Fortunately, he found a way to break the problem down into three one-dimensional problems that had already been solved. Drawing upon these solutions, which bore such esoteric names as *confluent hypergeometric functions* and *spherical harmonics* (the latter term has always struck the author as a bit poetic), Schrödinger was able to give an exact picture of the standing electron waves that replace Bohr's orbits. Several of these corresponding to some of the smaller Bohr orbits are shown in Figure 16-4. Where the pattern is brightest, the wave has its greatest amplitude.

By themselves, these patterns were impressive but not terribly informative. What *really* mattered was the formula for the energy levels that went with them. Schrödinger had met the supreme test: *the formula was exactly the same as Bohr's!*

Here at last was a complete theory, free from the ad hoc postulates of Bohr. Any force, any situation whatsoever was covered by it, without any need for additional assumptions or arbitrary rules. It also clarified the meaning of Planck's constant: it is the link between a particle's energy and momentum and a wave's frequency and momentum. Finally, it provided a way to visualize what was happening on the atomic scale, one that was intuitively appealing to those who were not experts in this area.

The wave picture of the hydrogen atom did more than remove the arbitrariness of the Bohr orbits. It also reconciled the quantum theory with Maxwell's, and eliminated the vexing instantaneous quantum jump. Applying the Maxwell theory to the wave patterns in Figure 16-4 showed that they should *not* radiate light. The transition from one pattern to another becomes a continuous process much like a motion picture "dissolve." One pattern fades

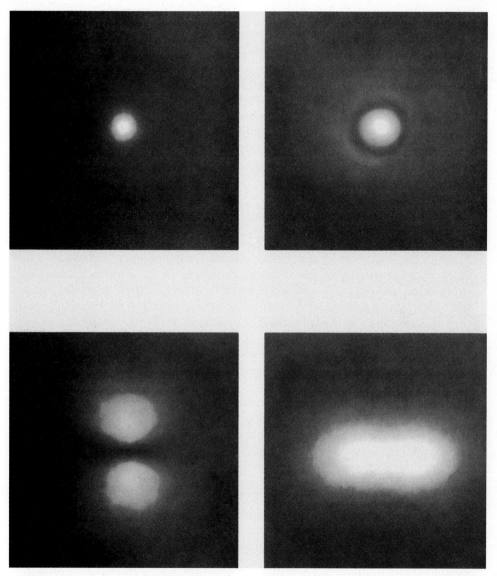

FIGURE 16-4. Examples of Schrödinger wave solutions for the electron in the hydrogen atom. The upper left picture corresponds to Bohr's original ground state: the others correspond to higher orbits.
(Reproduced by permission from J. Orear, Fundamental Physics, *2d ed., John Wiley & Sons, Inc., New York, 1967).*

out, while the new state fades in. During this time, interference between the two patterns produces oscillations that generate light of exactly the right frequency. The whole process typically takes around 10^{-8} second, a long time on the atomic scale.

To Einstein and Planck, disturbed by the irrational features of quantum physics, Schrödinger appeared as a savior. At last the accursed quantum jump had been replaced by a smoothly shifting wave pattern. It was Schrödinger's shining moment, and led to his most glittering job-hop. Planck retired in 1927, and Schrödinger was named his successor in Berlin. In 1933 he was awarded the Nobel Prize for wave mechanics.

THE PAULI PRINCIPLE AND CHEMISTRY

One of the most persuasive triumphs of the quantum theory of atomic structure was the way in which it provided a natural explanation of the regularities in chemical properties of the elements expressed in the periodic table. This explanation was originally worked out in the context of the Bohr model, but is most naturally expressed in the context of Schrödinger's wave model. By the time this explanation was proposed, the periodic table had been around for half a century, and represented one of the outstanding mysteries of science. Thus it was the solution of this problem that persuaded scientists outside the realm of atomic physics that the quantum theory was a valuable addition to science.

The first step toward a quantum theory of chemistry was the realization that the atomic number Z of an element, which was originally simply its sequence number in the periodic table, represents the number of electrons in an atom, which is equal to the number of positive charges in the nucleus.

The *Pauli Exclusion Principle,* an early contribution of Wolfgang Pauli to the quantum theory, is the key to understanding atomic (also nuclear) structure. It states that electrons are "territorial." To be specific, *no more than one electron can occupy a quantum state.* Each Schrödinger wave actually stands for *two* electron states, because the electron has an internal property named (somewhat misleadingly) *spin,* which has two allowed values. Spin corresponds to an angular momentum of $\frac{1}{2}\hbar$ Though electrons are often visualized as literally spinning on some axis, like many things in the microworld this simple visualization can be misleading. But roughly speaking, the two spin states correspond to clockwise and counterclockwise rotations.

Particles that obey the Pauli principle, which also applies to the particles that make up the nucleus, have a very special role in nature: they are the enduring "building blocks" of matter. This class of particles is known as *fermions.* We shall learn more about what makes them special in Chapter 19.

In the wave model, as in the modified Bohr model mentioned at the beginning of this chapter, each value of n corresponds not to just one state, but to n^2

different states. While for Bohr the additional states were arbitrarily added by allowing elliptical orbits, for Schrödinger they arose naturally from the allowed standing-wave patterns. All patterns with the same value of n are very nearly equal in energy. Electrons with the highest n values are also, on the average, much farther from the nucleus, so each set of electrons that share the same n is called a "shell." If we take into account spin, the nth shell is allowed $2n^2$ members, leading to the sequence $(2, 8, 18, . . .)$. Though the additional forces present in many-electron atoms disrupt the energy levels, all electrons in a shell tend to have roughly the same energy. Beyond $n = 2$ the shells tend to further subdivide into "subshells."

Atoms form chemical bonds through the interactions of the outer electrons, the ones in partially filled shells or subshells. In some cases, an electron actually moves from one atom to another, leaving two charged atoms (ions) that stick together by electrical attraction. This is called an *ionic bond*. In others, electrons take on complex patterns surrounding two or more nuclei, *covalent bonds*.

Only the electrons in the outermost shell or subshell participate in chemical binding, since it takes much less energy to remove them from the atom. These electrons are called *valence* electrons. Atoms with equal numbers of electrons in the outermost shell are similar in how they bond—this is the source of the regularity expressed in the *periodic table*. For example, hydrogen, lithium, and sodium each have one electron in their outermost shells, which are the $n = 1, 2,$ and 3 shells, respectively. All have similar chemical behavior, i.e., one can replace the other in most molecules. Subshells that contain eight electrons have a dominant role in atomic structure, which is why the periodic table has eight columns.

Elements with only a few valence electrons tend to give them up easily, as they are loosely bound. Similarly, elements with a nearly filled shell or subshell have room for an extra electron or two. Thus oxygen, which has six electrons in the $n = 2$ shell, can accept two electrons from two hydrogen atoms to form a molecule of water, H_2O.

Atoms with completely filled outer shells—the so-called noble gases helium, neon, argon, etc.—are essentially chemically inert because it is not energetically favorable for them to either donate or accept electrons.

Atoms with four valence electrons, e.g., carbon, silicon, germanium, are the most chemically versatile, for each can form as many as four chemical bonds at the same time. With four bonds available, carbon can serve as the backbone for complicated molecules containing many atoms, much as the "spools" in a tinkertoy set make complicated structures possible. The large molecules formed with the aid of carbon are the basis for life.

Protein molecules, for example, are a major constituent of living things, and some are built of more than 10,000 atoms. The pattern for building these molecules is recorded on *nucleic acids*, RNA or DNA, which can string together billions of atoms. Without four-valent atoms, no such molecules could be built.

WHERE IS THE ELECTRON?

Though what the Schrödinger equation had done was remarkable, there was one important thing that it did *not* do. It shed no light on the connection between the particle and the wave. In some ways, it even made matters *worse*.

For bound electrons, Schrödinger was content to view his standing waves as representing the electron itself. Though each wave filled the volume of the atom, that was still reasonably small. It could, with some stretch of the imagination, still be considered a particle.

But there was a more serious problem with the Schrödinger solution for the simplest situation in mechanics, a particle moving at constant velocity free from the influence of any force. The particle is represented as a *wave packet*, shown in Figure 16-5, a bundle of waves confined to a small region in space. But a wave is not a particle, and this packet refused to stay small. Like the wake of a boat, which spreads out from being a single sharp bow wave to a whole train of waves, the wave packet would spread out and spread fast. If initially confined to a space the size of an atom, it would spread in a fraction of a second to the size of the Great Pyramid! While Schrödinger's imagination might accept an electron the size of an atom, he could hardly reconcile himself to one as big as a pyramid. The problem clearly had to be resolved. And its resolution touched off the greatest controversy in physics since the time of Galileo.

FIGURE 16-5.

Wave "packet" representing free electron

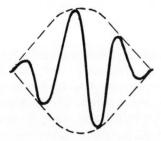

Same packet a very short time later

Summary

The key to moving beyond Bohr's model of the atom came from someone outside the mainstream of quantum physics, Prince Louis de Broglie. While restrictions on allowed patterns of motion were unnatural for particles, they were normal for standing waves. If Einstein had shown light could be both a particle and a wave, why couldn't the same hold true for electrons? A formula for the wavelength confirmed that Bohr orbits could be standing waves. Einstein promulgated de Broglie's idea among German-speaking physicists. One of these, Erwin Schrödinger, developed de Broglie's idea into a full-fledged three-dimensional wave equation that was fully as general as Newton's laws. His model for the hydrogen atom duplicated Bohr's energy levels but eliminated both the orbits and the quantum jumps. With the aid of a rule called the Pauli Exclusion Principle, the quantum theory of atomic structure provided a satisfying explanation of the chemists' periodic table of the elements, a triumph that helped the quantum theory gain acceptance in the scientific community at large. Still, the connection between the wave and the particle remained a mystery.

CHAPTER 17

Does God Play Dice?

But you tell me of an invisible planetary system where electrons gravitate around a nucleus. You explain this to me with an image. I realize then that you have been reduced to poetry: I shall never know. Have I the time to become indignant? You have changed theories. So that science that was to teach me everything ends up in a hypothesis, that lucidity founders in metaphor, that uncertainty is resolved in a work of art.

—ALBERT CAMUS, The Myth of Sisyphus

The Bohr model cast microphysics loose from the long Newtonian tradition. But the interpretation of Schrödinger's wave proved an even more serious break with history. In one stroke, it banished determinism completely. No longer could the future be predicted, even in principle. No degree of care in measurement could stay the inexorable hand of chance. Physicists could still speak of motion, but could no longer imagine that an object moves by orderly progress along a path. Reality on the atomic scale was reduced to a cosmic game of dice.

The formulation of this viewpoint took little more than a year from the publication of Schrödinger's theory. It took place in three main steps. First, Max Born saw a *probabilistic* connection between the wave and the particle it represented. Then Werner Heisenberg showed that any observation leads to an unpredictable change in the future of whatever is being observed. Finally, Niels Bohr interpreted this unprecedented situation as the revelation that science could no longer pretend to describe an objective "reality" in the traditional sense of the word. This chapter will outline the first two developments.

MAX BORN'S DICE GAME

Like Schrödinger, Max Born in 1926 was an established theorist who was facing middle age without a really big, original achievement to his name. But unlike Schrödinger, he was a man of modest ambition, content with his status in the profession. Everyone knew that the graduate students who came out of his institute at Göttingen were soundly trained. And he was happy to be of service to experimenters. It was this willingness to serve that brought him face-to-face with the wave-particle problem.

As Born explained it, "My Institute and that of James Franck were housed in the same building. . . . Every experiment by Franck and his assistants on electron collisions . . . appeared to me as a new proof of the corpuscular nature of the electron." Franck was following in the tradition of Rutherford, but using electrons rather than alphas as probes for the structure of matter. His experiments revealed without question that the electron was a particle. One could even follow an electron in a Wilson cloud chamber, where it left a trail of water droplets to mark its path. It certainly did not blow away into nothingness, like Schrödinger's wave packet.

Born set out to develop an analysis of particle scattering based on wave mechanics. Rutherford's formula for the scattering of alphas, the inverse fourth power of the sine of half the angle of deflection, was by then well-tested. Born knew that one sure way to check the soundness of his calculations would be to reproduce that result. He visualized an alpha particle as a series of waves approaching a nucleus, as illustrated in Figure 17-1. With the aid of the Schrödinger equation, he asked what the encounter with the nucleus would do to these waves. As he expected, most of the wave amplitude

FIGURE 17-1. Wave representation of Rutherford scattering.

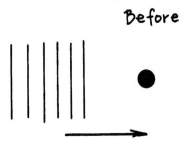

Before

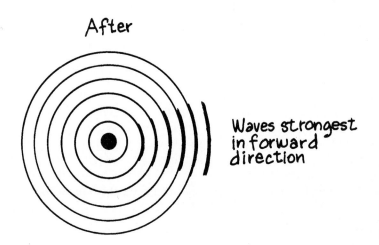

After

Waves strongest in forward direction

emerged unchanged. But a small portion was reflected, in a pattern radiating out from the nucleus in all directions.

But the wave was not *equally strong* in all directions. In fact, its amplitude was proportional to the inverse *square* of the sine of the half-angle! Rutherford's formula was almost there before his eyes, but all he had was a wave. Where was the alpha particle? He realized that to get from his wave picture to Rutherford's formula he needed a quantitative connection between the wave and the particle. And the only connection that would do the trick was through a statement about *probability:*

> The probability of finding the particle in some particular place* is proportional to the *square of the amplitude* of Schrödinger's wave at that location.

Since the square of a squared quantity is its fourth power, this assumption connected Born's wave amplitude to Rutherford's probability that an alpha particle will be deflected at some particular angle. Rutherford's formula gave only a probability because the experimenter had no way of knowing how close a particular alpha would come to a particular nucleus. The orbits themselves were perfectly deterministic. Born, however, had enshrined probability at the *heart* of physical reality: if Schrödinger's wave described what was happening, then *only* probabilistic predictions could ever be made!

So the wave does not tell the physicist where a particle *is* at any given moment, but merely where it is *likely to be*. For example, one could measure the position of an electron in the ground state of an atom as precisely as the instruments will allow. Repeating this measurement many hundreds of times and plotting the results in the form of dots on a picture, as in Figure 17-2, would give a pattern of dots resembling the wave patterns shown in the preceding chapter. No individual measurement can be predicted with any greater precision than to say it will fall somewhere in the pattern.

Determinism itself had been abandoned. No longer could a measurement be used to predict the exact future. The wave itself is perfectly well-behaved and predictable in its development, but it has only a random connection with observable reality. We can still predict the average of a large number of repeated measurements, but the result of any individual measurement must forever remain a surprise.

Probabilistic laws were no novelty in physics—they were the stock in trade of statistical mechanics. There one always spoke of average speeds of atoms, or average distances between collisions, without trying to trace the motions of the individual atoms in detail. This merely reflected practical ignorance, the impossibility of coping precisely with the motion of 10^{23} incredibly tiny atoms as individual objects. No one doubted that the details of the motion were subject to Newton's laws, and if one were given the staggering amount of information required, one could exactly describe the future motion of every atom. What Born seemed to be saying was that there was no way whatsoever to predict the precise future position of even one isolated atom.

---

*To be exact, it is the *probability per unit volume,* because the chances of finding the particle depend on how big the region is where you search for it.

Schrödinger.
(Photo by Ullstein, courtesy of American Institute of Physics.)

FIGURE 17-2. Results of repeated measurements of
the position of an electron in the first Bohr orbit.

UNCERTAINTY

Schrödinger's theory was welcomed in Copenhagen, especially after Born's
probability interpretation made it clear that it was not the beginning of a
retreat back to the old physics. But the same intuitive clarity that made it popular in the broader community made it somewhat suspect to Bohr. In the
world of the atom, Bohr was sure, visualization could only lead one astray.
Better to deal in abstractions, and not be tricked by images and words.

Working closely with the Copenhagen physicists, Werner Heisenberg, a
postdoctoral fellow at Göttingen, had developed his own approach to the
quantum theory. He never mentioned *either* particles or waves, but spoke only
of quantum *states*. Even the name of this theory proclaimed its austere neutrality: it was called *matrix mechanics*, for the abstract algebraic forms (matri-
ces) in which it was expressed.

Without intuition as a guide, Heisenberg was unable to get much mileage
from his theory. Only after Schrödinger showed the way did Pauli succeed in
solving the hydrogen problem by Heisenberg's methods, and it was soon discovered that the theory's abstract "states" were in fact solutions to the
Schrödinger equation. Nonetheless, Heisenberg's formulation is favored by
most physicists to this day, because of its mathematical elegance.

After Born had shown that it was unlikely that any version of the quantum theory would ever give deterministic answers, Heisenberg explored his
abstract algebra for a clue to the origins of this lack of definiteness. He soon

found a rule that set quantitative limits to our ability to know what is going on in the microworld. He decided that this indeterminacy was indeed the result of ignorance, but ignorance of an inherent and unavoidable kind. His interpretation was that the indeterminacy arose from *the disturbance of an object in the act of observing it.* Today we have come to realize that Heisenberg's rules are far more general than that, and apply to situations in which no measurement, in the usual sense, is involved.

The specific rules Heisenberg developed to quantify the degree of disturbance are called *uncertainty relations.* They state that there are certain *pairs* of physical quantities that cannot be determined simultaneously to any desired accuracy. The measurements themselves may be as accurate as the instrument maker's skill allows, but repeated measurements will not exactly agree, and a certain error must be assumed when using such a measurement to predict the future.

One such pair of variables is *position and momentum,* and another is energy and time. The uncertainty relations state that the *product* of the uncertainties of the two variables may not go below a certain minimum value, which turns out to be $\frac{1}{2}\hbar$. That means that if one of the variables is well-determined, the other must be assigned a larger error to make the product of the two uncertainties come out large enough.*

An error in momentum must mean an error in velocity, since mass is usually well determined. And it is precisely the simultaneous knowledge of position and velocity that is essential to knowing where an object will be in the future. If we know where an object is but have only a rough estimate of its velocity, it is hard to say where it will be some time later. But if we know how fast it is going but are uncertain where it is now, we are just as badly off.

Heisenberg derived his law on quite general and abstract grounds. Its true significance becomes apparent only when one shows how it enters into any specific situation. This is another *gedanken* experiment game, like those we played in the development of relativity. The law works differently in every imaginable process, but its consequences are inescapable.

In a measurement of the position of an electron, at least one photon must interact with the electron and then be detected. The measurement cannot be much more accurate than the wavelength of the photons employed. But a short wavelength means a high momentum, causing the electron to recoil from the photon, changing its momentum.

The significance of the energy-time uncertainty relation is somewhat different. It implies that in order to have a well-defined energy, a physical state must last for a long time. The energy of a short-lived excited atomic state is slightly "smeared out," as revealed by a good spectrograph. The lines are not perfectly sharp, but each has a natural width that depends on how long the

---

*For readers familiar with statistics, Heisenberg's rules relate the standard deviations, e.g., δp and δx for momentum and position: $\delta p \delta x \geq \frac{1}{2}\hbar$.

state it came from survived. The ground state energy is well defined because this state can last as long as the atom remains undisturbed.

The *momentum* of the electron in the hydrogen ground state, however, is *not* well-defined. We have no idea of what direction it is moving in at any given time, or of how its total energy divides up between kinetic and potential energy. The spread in space of the standing wave represents our uncertainty as to its position. The product of uncertainty in momentum and position exactly match the lower limit set by Heisenberg.

In the final section of this chapter, we will see that the indefiniteness demanded by the Heisenberg relations is terribly significant for the way the microworld functions. It permits things to happen that would be quite impossible in Newtonian physics.

With the formulation of the uncertainty relations, Heisenberg could at long last make his life all of one piece, and hold up his head before his Youth Movement companions. In a popular article for the *Jügendbewegung* newspaper Heisenberg proclaimed the demise of determinism, struck down by the work of dedicated young German physicists!

THE "OLD MEN" WON'T BUY IT

The work of Born and Heisenberg brought shock and dismay to Einstein, Planck, de Broglie, and Schrödinger. They had seen order and continuity restored to the microworld, only to see it snatched away a few short months later. Einstein in particular simply refused at first to accept the validity of the uncertainty principle, a position summarized in a celebrated remark to Bohr: "God does not play dice!" Bohr replied, "Einstein, stop telling God what He can and cannot do!"

Einstein's first reaction was to search for counterexamples, *gedanken* measuring procedures that would be exempt from the principle. But these proved as futile as the similar attempts made earlier against his own relativity theory. The others gradually and reluctantly came to accept the new view.

A celebrated visit by Schrödinger to Bohr's institute in the fall of 1926 was the turning point. After days of debate lasting well into the night, Schrödinger finally conceded defeat with the outburst: "If one has to stick to this damned quantum jumping, then I regret ever having gotten involved!"

THE PROBLEM OF PREDICTION

The degree to which an uncertainty relation restricts our ability to predict the future position of an object depends on its mass. The relation limits our knowledge of *momentum*, while it is *velocity* that is used to predict the future position. The more mass an object has, the less important the uncertainty in its momentum becomes, because it corresponds to a smaller uncertainty in velocity.*

Objects big enough to be visible to the naked eye have large masses by atomic standards. On this scale, errors in position as large as a light wavelength are inconsequential, so a fairly large δx is tolerable. Thus the uncertainty in momentum is small to begin with, and when we divide it by mass to get the uncertainty in velocity we find it is completely negligible. We can use Newtonian physics in the world of everyday experience with complete confidence. On the atomic scale, masses are smaller and the required precision is greater, so prediction becomes nearly impossible.

This is illustrated quantitatively in Table 17-1, which shows how far in the future we can make useful predictions, working at various scales of size. It was obtained in the following fashion. In the first column we indicate the accuracy of length measurement appropriate to a particular scale of size. In the second, we show the mass of a typical object found on that scale. The third column gives the uncertainty in velocity of the object, calculated from the uncertainty relation for position and momentum. The fourth column tells us how long it will be before the uncertainty in velocity will lead to a position error larger than that specified in the first column. After this time, the uncertainty in velocity dominates, and our predictions get worse and worse.

The first example, the electron, illustrates the "expansion of the wave packet" described at the end of Chapter 16. With a velocity uncertainty of 10 million meters per second, the electron could be anywhere in a volume the size of the Great Pyramid in a terribly short time. On the atomic scale, predictions are good for roughly the time between atomic collisions in a gas. Biochemical processes are predictable on a time scale of milliseconds, in which some fairly significant things can take place.

*Rewriting the uncertainty relation as a relation between errors in position and velocity, we get $\delta x \delta v \geq \hbar/2m$. A large mass in the denominator makes the uncertainty small. This formula is the basis for Table 17-1.

TABLE 17-1

Scale, in Meters	Object and Mass, kg	Uncertainty in Velocity, m/s	Prediction Time Limit, Seconds
Subatomic 10^{-1}	Electron 10^{-30}	10^7	10^{-18}
Atomic 10^{-10}	Atom 10^{-25}	10	10^{-11}
Biochemical 10^{-8}	Big molecule 10^{-21}	10^{-5}	10^{-3}
Microscopic 10^{-6}	Pollen 10^{-15}	10^{-13}	10^7
Macroscopic 10^{-5}	Pea 10^{-4}	10^{-25}	10^{20}

Anything visible in a microscope, however, is pretty much Newtonian. Brown's pollen grains danced about not because of quantum mechanics, but because of the randomness of molecular impacts. The motion of a pea is supposedly predictable for the entire age of our universe, but it would have to be a terribly isolated pea, lost in intergalactic space, for this to really hold true. Even on these levels of reality, most of our predictions eventually go wrong for reasons having little to do with quantum effects, as the computer studies of *chaos* mentioned at the end of Chapter 6 forcefully remind us.

ATOMS LEAVE NO TRACKS

One of the more disturbing consequences of the quantum theory is that it does allow us to think of an electron in an atom as having any kind of path at all, much less a tidy Keplerian orbit. To do so would require us to observe the electron in one position, and then repeat the observation at short intervals, mapping out a path.

But each observation involves the exchange of at least one quantum with the electron. An electron in an atom is not free to interact with just any quantum, for it is not free to assume any energy. The very *least* reaction it can have is to be kicked into some higher energy state. Thus an electron can be observed only once in any given state. We must accept that if the electron did follow an orderly path in the atom, we would never be able to map it out.

Even worse, despite its lack of a well-defined path, the electron stubbornly retains all the other attributes of motion—momentum, velocity, and so on. It seems absurd to speak of motion without a path, but the quantum theory leaves us no other choice.

QUANTUM TUNNELING

Heisenberg interpreted his uncertainty relations in the context of the theory of measurement. Our final example in this chapter will concern an effect that shows the consequences of uncertainty are far broader. They can turn up in situations that do not directly involve measurements at all.

The effect is called *tunneling*, and it is impossible to overstate its practical importance. It is an essential mechanism for the nuclear reactions that power the Sun and other stars, as well as for nuclear fission. Much of modern microchip technology is based on tunneling. In Chapter 19, we shall see that it permits the key process by which all of the fundamental fields of force operate.

Though many physicists contributed to this idea, its clearest and most useful formulation came from the Bohr Institute's merriest prankster, the Russian émigré George Gamow, whom we have already mentioned in conjunction with the big bang cosmology, one of his later achievements. He arrived in Copenhagen in 1928 from Leningrad riding a motorcycle, an appropriate conveyance for a restless spirit who throughout his life steadfastly refused to settle down. He was an entertaining public lecturer, and authored a number of very successful (and sometimes hilarious) popular books.

But along with the highjinks came a prodigious flow of original ideas. Gamow was a pioneer in applying the quantum theory to the mysteries inside the nucleus. He is also the source of our modern understanding of supernovas, and even contributed one key insight to early speculation about the genetic code. Gamow's work on tunneling was done in a brief stopover at Cambridge University in the 1930s. Following a brief sojourn in his native land, he moved in 1934 to the United States.

Simply stated, the idea behind tunneling is that *the uncertainty relations allow you to get away with anything, as long as you do it fast enough!* In a process that happens rapidly, the energy is simply not well-defined. A particle may quickly enter and leave a state that it doesn't have enough energy to remain in. Put another way, a process that would be *impossible* in Newtonian physics is merely *improbable* in quantum physics.

Imagine a car coasting up a hill with the engine turned off. If it has too little speed to make it over the top, it will come to a stop and start to roll back. If it could somehow be miraculously transported to the same level on the other side of the hill, it could coast downhill and continue on its journey. In the microworld, this sort of thing can really happen, as long as it happens quickly enough.

For another kind of analogy, a quantum system can be something like a dishonest bank official who borrows money on the sly, but always manages to pay it back before the bank examiners can tell it is missing.

A classic example of tunneling is the nuclear fusion reactions that power the Sun. If two hydrogen nuclei can get close enough, a reaction can take place that fuses them together, and energy is released. But the strong mutual repulsion of nuclei keeps them apart: even at the temperatures found in the Sun's core, they do not have enough energy to overcome this barrier. But they can come close enough to give a small but significant probability that they will tunnel through and fuse. Since the nuclei collide frequently, before long a reaction does take place.

Tunneling is possible in any situation in which a system with a given amount of energy must pass through a state of higher energy in order to reach one with the same or lower energy, illustrated schematically in Figure 17-3. The intermediate state is like the hilltop in the example of the car. An object passing quickly through this forbidden state is said to be in a *virtual* state. The probability of tunneling is very sensitive to the difference in energy of the allowed and forbidden states, and to the time it takes to pass through. Gamow developed the formula that gives this probability.

Tunneling demonstrates that the uncertainty relations don't merely limit how well things can be *measured*—it shows that they limit how well they can be *defined*, whether measured or not. There is a kind of inescapable "fuzziness" to physical quantities in the quantum world. In Chapter 19, we will see that this fuzziness is in fact a welcome feature of the theory. Without it, we could not have *fields*, the very fabric of both matter and the universe.

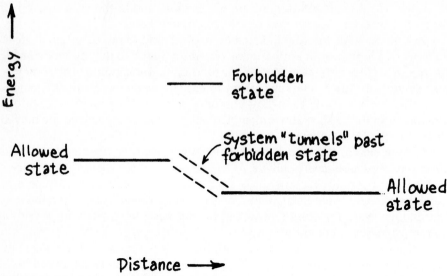

FIGURE 17-3. Quantum tunneling.

Summary

Through an analysis of particle collisions using Schrödinger's waves, Max Born concluded that the wave represents not the particle itself, but the probability of finding the particle in any particular place. Though the wave form can be predicted exactly, only statistical predictions can be made for the particle. Werner Heisenberg quantified this through his uncertainty relations, which set limits on the accuracy to which certain pairs of variables can be simultaneously determined. Heisenberg viewed this as a consequence of the disturbance of a system by the act of observing it, but the relations have more general significance. They limit our ability to predict the future, and give rise to a process called tunneling through which things that would be impossible in classical physics become simply improbable.

CHAPTER 18

Schrödinger's Cat

The law of chaos is the law of ideas,
Of improvisations and seasons of belief.
—WALLACE STEVENS,
Extracts from Addresses to the
Academy of Fine Ideas

The quantum theory had become a complete paradigm by 1927. Over the years since, the theory has been refined, applied, and extended, but none of this elaboration has in any way shaken the fundamental principles on which quantum physics rests. The probability interpretation and the uncertainty relations stand unchallenged.

But the debate over what it all *means* continues unabated to this day. This controversy has generated hundreds of books and thousands of articles. Physicists, philosophers, mystics, and even poets have had their say. Although a few naive ideas have been put to rest along the way, there is still no hint of a final consensus. In this chapter we will explore some of the more compelling or amusing viewpoints on this issue, through a series of examples.

The source of the controversy is that the quantum theory assigns only probabilities to *possible* outcomes of an experiment. Nowhere does the theory indicate that the experiment will, in fact, have one and only one *actual* outcome. Niels Bohr thought he had the answer to this problem: by the *act of measurement* we push nature into giving us one answer or another. Thus if we include the measuring procedure and apparatus in the description of a physical situation, all will be well.

Unfortunately, however, matters do not end there. When measurements are included, the theory still gives only probabilities, now assigned to all the possible *readings of the measuring instruments*. This led Bohr and others under his influence to a line of reasoning that came to be called the *Copenhagen Interpretation* of the quantum theory. To illustrate this interpretation, let us examine how it treats the "expanding wave" that describes a free particle.

THE WAVE COLLAPSES

The uncertainty relations show us the significance of the expanding wave packet. Its initial size represents the uncertainty in our knowledge of the position of the particle. The spread of the wave packet with time arises from the momentum uncertainty. Since there is some uncertainty as to how fast the particle is going, as time passes, the region in which it might be grows bigger and bigger.

When we look at the expansion in purely wave terms, the uncertainty in momentum corresponds to a spread in wavelengths. The wave packet is therefore a mixture of waves of varying wavelength. Since, in the case of the Schrödinger wave, waves of different length travel at different speeds, the packet will naturally expand.

If we make a number of successive observations, the Copenhagen description of the wave, illustrated in Figure 18-1, takes on a peculiar new twist. At each observation the wave "collapses." We start all over again with a new and smaller wave packet, which again expands until the next measurement ties down the electron's position again. This collapse does not come out of the quantum theory itself: it must be *imposed* on the theory to represent the effects of the measurement, which converts the *possibilities* that the wave represents to the *actualities* of instrument readings. But since the theory only assigns proba-

FIGURE 18-1.

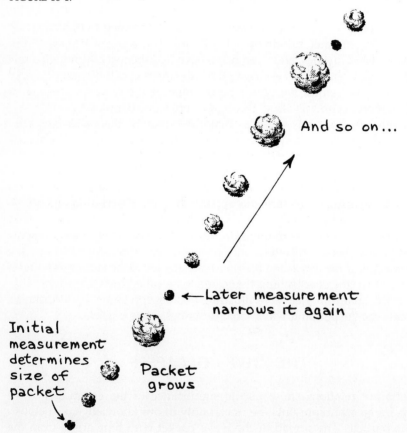

And so on...

Later measurement
narrows it again

Initial
measurement
determines
size of
packet

Packet
grows

bilities to these readings, when does the collapse take place? The Copenhagen answer is that it occurs when the experimenter *becomes aware of the result of the measurement!*

In the Copenhagen view, the Schrödinger wave does not represent the *particle itself* but *what we know* about it. Through the quantum theory, nature teaches us a lesson in philosophy. We only *pretend* that science is about nature itself—in reality, science can only depict what we *know* about nature, and in a probabilistic theory, we can't know everything.

After his celebrated confrontation with Bohr, Schrödinger had become more favorably disposed toward Born's probability interpretation of the wave, at least as a working hypothesis. But this business of dragging the experimenter's *mind* into physics was, for him, the last straw. After a few years of brooding, he struck back in 1935 with a parable that has come to symbolize the whole debate.

THE PARABLE OF SCHRÖDINGER'S CAT

Schrödinger's assault on the Copenhagen Interpretation took the form of what he called a "quite burlesque case," which may be paraphrased as follows:

> A cat is placed in a sealed box. The box is equipped with a diabolical apparatus, triggered by a Geiger counter that contains a few radioactive atoms. The device will release a lethal gas that kills the cat. There is a 50 percent probability that the mechanism will trigger in any given hour. At the end of one hour, a physicist opens the box and finds out whether the cat is alive or dead.

In this situation, the Copenhagen Interpretation regards the cat as simply a furry component of the measuring apparatus. It is a link in the chain leading from the microscopic quantum indeterminacy of a radioactive atom to the concrete macroscopic image of a live or dead cat in the experimenter's mind. Since the "wave function" that describes the system must include the state of the entire measuring apparatus, an instant before the box is opened it contains an equal admixture of "live cat" and "dead cat." The cat's fate is settled only when awareness comes to the observer's mind, at which time the pattern collapses into one state or the other.

In Schrödinger's view, the experimenter's mind is just another part of physical reality, with no particular privileged status. Why doesn't the *cat's* mind rule the collapse of the wave function? For that matter, what happens if we replace the cat with a human being or, at the other extreme, a mechanism that stamps the time on a slip of paper? Why doesn't the wave collapse *then?* The Copenhagen Interpretation would burden science with the "mind-body distinction," the notion that human consciousness has a special character above and beyond the physical brain that harbors it. Schrödinger had no objection to this view, but insisted that it had no legitimate place in physics.

For Bohr himself, the parable posed no problems. Since science is about what we *know*, he was perfectly happy to let the cat and the experimenter know different things. That "half dead, half alive" wave pattern, and its "collapse" into one outcome or the other, are simply mathematical artifacts, parts of the calculus we use to estimate the probability that the cat will live or die. They should not be taken as representing the *reality* of the situation, which he regarded as essentially unfathomable by minds attuned to the realities of our macroscopic world.

But some of Bohr's more ardent followers, notably John von Neumann, Eugene Wigner, and John Wheeler, actually welcomed the situation. They were perfectly happy to give the human mind a special role in the universe. Wheeler calls this a *participatory* role: the human consciousness actually *creates* reality by observing it! The big bang created our universe in some sense *because* ten billion years later there would be human beings with minds that could decipher the clues that point back to the cosmic explosion. *No mind, no universe.*

This view goes way beyond Bohr's, and should probably be distinguished from the Copenhagen Interpretation. Since its three most celebrated supporters spent a major part of their careers in Princeton, New Jersey, it might be more appropriate to call it the *Princeton Interpretation.*

It would have seemed quite natural for Schrödinger to embrace this view with enthusiasm. In his inaugural lecture at Zurich in 1922, he had pointed out that the assumption that atoms follow deterministic laws was unnecessary and possibly false. And he was an adherent of Vedantic philosophy, with its "world soul," a cosmic intelligence shared by all things living and nonliving. To many adherents of the Princeton Interpretation, the quantum theory moves physics toward this worldview.

Most physicists have little use for speculative philosophy, dismissing it as a "soft" discipline in which arguments are never definitively settled. Some, like Bohr, value philosophy and see physics as a route to new philosophical insights. Schrödinger felt that physics was a lower form of knowledge that should stick to the mundane world and not pretend to shed light on deeper questions. Thus he rejected an interpretation of the quantum theory that seemed in harmony with his own philosophical views.

There are even more bizarre interpretations of the quantum theory. One, also associated with Wheeler and some of his students, is the *many-worlds* hypothesis. In this vision, Schrödinger's cat *both lives and dies.* Somehow, our minds are aware only of one part of this vast reality, which includes all the things that *might* have happened in the entire history of our universe! Thus true reality consists of a myriad of parallel universes, in half of which the cat lives and in the other half it dies. Whether all our minds perceive the *same universe remains an open question.*

The most unorthodox school is the "physics and consciousness" movement, of which the principal spokesman is Jack Sarfatti. They embrace the many-worlds hypothesis and see in it the basis for such "occult" phenomena as extrasensory perception (ESP). If the experimenter cultivates his or her psychic powers, perhaps he or she can consciously *will* the fate of the cat! In honor of the domicile of most adherents to this viewpoint, we might call it the *California* (where else?) *Interpretation.*

At the other extreme, Einstein remained convinced to his dying day that the cat's fate was sealed when it went into the box. If the quantum theory can't predict it, then it is an incomplete theory that will someday be supplanted. Motivated by deep mutual respect, Einstein and Bohr fought this battle for twenty-five years, each failing to budge the other from his point of view.

Einstein pointed out that when two objects interact and then move apart, conservation laws enable us to tell such things as the momentum of one object from measurements on the other, even though they are then too far apart to influence one another. This convinced Einstein that the outcome of the measurement must somehow have been predetermined while the objects were still in contact, rather than at the time of measurement as the Copenhagen school insisted. The paper outlining this view was written with two young colleagues of Einstein, Boris Podolsky and Nathan Rosen, so it goes by the name *EPR Paradox*. But this position turns out to be one of those that has been effectively eliminated from contention in recent years.

In 1964 John Bell, an Irish theorist working at an international laboratory in Switzerland, proved a theorem that applies to experiments similar to those cited by Einstein. In Bell's case, it is the *spins* of particles that have moved far apart that are measured. He showed that if we simply assume that no measurement can influence another unless they are connected by either a common cause or a causal link that is transmitted no faster than the speed of light, we calculate correlations between the spin measurements that differ from those predicted by the quantum theory. Such experiments were performed in the 1980s, and the quantum predictions won out. While the quantum theory may someday be supplanted, it seems highly unlikely that any successor paradigm will restore simple Newtonian determinism.

Bell's theorem has deeper philosophical implications. It appears to rule out the view that reality can be separated into the sum of individual realities found in different locations in space and time. It suggests a "wholeness" to reality that can extend over vast regions of the universe. This feature can be summarized as follows: "reality is nonlocal." The meaning of this statement will become clearer in the final example in this chapter.

There are many other positions on the implications of the quantum theory. One favored by some philosophers would require us to abandon our usual notions of logic, in which statements are either "true" or "false," for a new "quantum logic" in which they are simply "probable" or "improbable." But we will close this discussion by presenting two views perhaps best described as *agnostic*.

Leslie Ballentine of Canada feels that the whole problem arises from the fallacy of using a *statistical* theory to describe *individual* events. By their nature, statistical predictions can be confirmed only by a large number of trials. In his view, the "half-live, half-dead" cat simply represents the fact that if we repeated the experiment 1000 times, the most probable outcome would be 500 live and 500 dead cats.

Einstein had advanced the same argument, but he insisted that it proved that quantum mechanics was an *incomplete* theory. Ballentine is willing to accept that the most fundamental laws of science may not enable us to completely account for every individual event that happens in our universe.

We will give the final word in this debate to Richard Feynman, an American whose personality and ideas dominated theoretical physics for much of the latter half of the twentieth century. He will be the principal hero of the final chapter of this book. Though revered within his profession, Feynman was little-known to the general public until shortly before his death in 1989, when he served on the panel investigating the fatal accident of the space shuttle *Challenger*. There, with the aid of a glass of ice water, he demonstrated on live television the rigidity of the rubber seal that had led to such tragic consequences. The stunt was vintage Feynman, both in its simplicity and its flair for showmanship.

The hallmark of Feynman's style, like that of Bohr, was *irreverence*. But, unlike Bohr, he extended this principle to the philosophical ruminations of distinguished scientists. In his view, researchers are merely "kibitzers" observing the "game" of nature and trying to figure out the "rules" by which it is played. It is hard enough to figure out the rules without the help of a rule book—it is sheer arrogance to insist that the rules should also *make sense*. He regarded thinking about the deeper significance of the quantum theory as a blind alley, at the end of which lies not enlightenment but madness.

It was more constructive, in Feynman's view, to try to define the essence of how the quantum rules differ from those that applied in Newton's universe. We could then admire these rules for whatever aesthetic qualities they might possess, but in the long run we can only humbly accept that they are what they are, forever inexplicable. And he felt that the essential quality was best revealed in the very experiment that established the wave nature of light.

YOUNG'S EXPERIMENT WITH ELECTRONS

Young's experiment, the interference of light passing through two closely spaced slits, was introduced in Chapter 7 as a wave phenomenon. To emphasize the particle aspects of the effect, let us imagine it to be performed with electrons, because the electric charge they carry makes it possible to observe them in flight. This is not possible with photons, which can only be created or absorbed. The Schrödinger wave represents each electron individually. At a reasonable beam intensity, only one wave packet moves through the apparatus at a time.

If we view the electron as a wave, our analysis of the experiment is precisely the same as Young's. On a viewing screen behind the slits, there are places where waves arrive crest to crest. There the wave is strong, and many electrons are observed. Where the waves meet trough to crest and cancel, few electrons appear. The resulting pattern of light and dark bands is shown in Figure 18-2. The only change introduced by quantum mechanics is that instead of a continuous glow, there is a series of flashes.

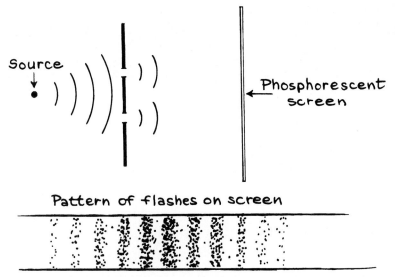

FIGURE 18-2. Young's experiment with electrons.

But from the particle point of view, the result is most perplexing. Surely each electron that reaches the screen must have passed through one or another of the slits. It is hard to imagine that the presence or absence of the other slit, the one it didn't go through, can have any influence on where it hits the screen. Nonetheless, if we close one of the slits the interference pattern will disappear, to be replaced by the blurred image produced by a wave that diffracts through the slit. If each electron passes through just one slit, why don't we simply get two overlapping single-slit images, as shown in Figure 18-3?

Does each electron, as it passes through one slit, somehow "know" whether the other slit is open or closed? The answer, of course, is that the electron wave describing a single particle does pass through both slits. But this merely deepens the mystery. What then does this schizophrenic wave have to do with a particle? The electron left its source in one piece, and wound up at a particular location on a screen, where it made one tiny flash. And with a simple modification of the apparatus, we can even tell which slit each electron went through.

This modification is shown in Figure 18-4. It is the same as before, except we now have a loop of wire around each slit to sense the magnetic field that

FIGURE 18-3. Pattern of flashes from two slits never open at the same time.

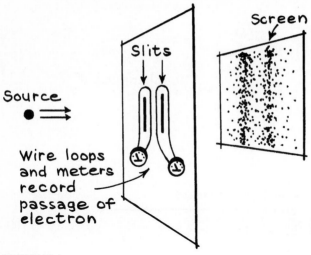

FIGURE 18-4.

the electron generates as it passes. Observing the electrons one by one, we find that only one loop at a time will give a signal. The particle indeed passes through one hole or the other. But that is not all we discover. On the screen, the interference bands have vanished! The pattern is that of Figure 18-3.

We have added another measurement into the experiment, and the effects of this measurement must be taken into account. To generate a signal in the loop of wire, the electron had to emit or absorb at least one photon. This changed its momentum and wavelength by an unknown amount, destroying the synchronization with the wave from the other slit.

In Feynman's view, this example pinpoints the distinction between quantum physics and Newtonian physics. In Newton's world, the electron would move from its source to its destination along a well-defined path, and only the forces it encounters along that path affect its destiny. In the quantum theory, not only is the electron's path indeterminate, but the probability of reaching any particular destination depends on what happens on all of its possible paths. If we reduce the number of possibilities, either by obstruction or by observation, we simplify the situation and the interference effect disappears.

No longer can we regard the present as the final consequence of a single, unbreakable chain of events. Instead, we must consider all the "undismissed possible pasts" that *might* have contributed to the present. Causation ceases to be *particular* and becomes somewhat *holistic*. This is another example of the "non-locality" implied by Bell's theorem. Since it is the nature of waves that each part of the wave can influence the future of all other parts of a wave, it is in this sense that a particle is "wave-like."

As reasonable as these arguments may seem, many physicists and philosophers feel that Ballentine and Feynman are passing up an opportunity for deeper insights. Thus the debate on the meaning of the quantum theory is likely to persist as long as the theory itself.

But a working physicist need not be bothered by these questions. It is enough to know the rules, and to use them with skill. Thus employed, the quantum theory has proved to be a fine scalpel, allowing scientists to probe the very heart of matter. What they have found there will be the topic of the final chapter.

Summary

Faced with the philosophical problem of connecting a probability wave with the reality that consists of actual individual events, Bohr proposed that the act of observation turns the many possibilities into a single actuality, a position called the *Copenhagen Interpretation*. Schrödinger attacked this view through his "cat parable," which can serve as a vehicle for presenting a variety of interpretations of the quantum theory. The debate rages to this day, though a simple deterministic interpretation is now definitely ruled out by measurements suggested by John Bell. The lack of resolution of these issues has no effect on the utility of the theory, and some physicists, notably Richard Feynman, have argued that the issue is a false one.

CHAPTER 19

The Dreams Stuff Is Made of

Like a gleam in the darkness, we have appeared for an instant from the black nothingness of the ever-unconscious matter, in order to make good the demands of Reason and create a life worthy of ourselves and of the Goal we only dimly perceive.

ANDREI SAKHAROV

Up to now, this narrative has been concerned with ideas that are many decades old. Like fine wine, age has mellowed them and taken away some of their initial roughness. In this concluding chapter we will explore more recent developments. Like a construction site, frontier science is never tidy. There is still a great deal of machinery and scaffolding lying about, and it sorely needs landscaping. In science, this means a plethora of confusing facts and baffling terminology. So it is important to keep in mind a central theme: *seemingly solid matter is nothing more than a manifestation of fields that do not "occupy" space at all.*

Modern physics has accepted Rudjer Boscovich's challenge, which was briefly touched upon in Chapter 6. The search for the ultimate particles of matter can end only with the discovery of structureless, point-like objects. But space is far from empty, because on a small enough scale the uncertainty relations allow little bundles of energy to pop in and out of existence. It is of such "virtual" objects that fields are made.

Unfortunately, the step to the level of reality we are about to explore did not bring with it the reductionists' hoped-for simplicity. All we can say is that Anaxagoras gave us fair warning: *deeper* levels of reality are not necessarily *simpler* levels. So this chapter will explore a realm that may seem dauntingly complicated.

The development of this picture began with the emergence of the modern quantum theory in 1927, and now engages the talents of thousands of experimenters and theorists around the globe, including the author of this book. Though remarkable discoveries have been made, the picture is by no means complete. Before presenting the most up-to-date version, which bears the rather uninspiring name of the *Standard Model*, let us review some of this history.

FAREWELL TO INNOCENCE

Though the quantum theory that emerged in 1927 was logically complete and self-consistent, the physicists who developed it knew enough about the history of their science to realize that their work was far from over. Like Newton, they had created a *kinematics*, a scheme for describing and predicting motion, without a full understanding of *dynamics*, the origins of the forces that govern that motion.

Newton's scheme had to wait nearly 200 years for its dynamics, the Law of Energy Conservation and field theory. The quantum physicists expected far more rapid progress, however, because they knew precisely what had to be done. They needed a quantum description of the *fields themselves*.

Schrödinger's treatment of the hydrogen atom had used Maxwell's classical electromagnetic field. But surely this could be only an approximate picture. The "free" electromagnetic field—light itself—was known to be quantized. Einstein's photons were now universally accepted. What remained was to bring photons into the theory of electric and magnetic fields, and then repeat this conversion for gravity and whatever other fields were needed to account for all the known forces.

The quantum theorists had expected to make short work of this problem, but they had underestimated the magnitude of the task. There are more fields in nature than they realized, and they are so closely related that it is not possible to fully understand one in isolation from the others. And far too little was known about the subatomic particles themselves.

They soon ran out of time. In 1933, Europe's golden age of "modernism" came to a sudden and violent end. Within months of Hitler's ascent to power the German scientific community was in shambles. Nearly all Jewish professors had been dismissed, and though they were few in number they were concentrated in frontier areas of research. A few liberals followed them into exile. Einstein had been fortunate enough to be on his way to what had been planned as a six-month stay at the newly created Institute for Advanced Study in Princeton, New Jersey, when the takeover occurred, for the new regime immediately put a price on his head. Princeton became his home for the remainder of his life.

Though few Germans realized it at the time, the future of German science had been dealt a blow from which it has not fully recovered to this day. Even Schrödinger, though a gentile and haughtily disdainful of politics, soon left his prestigious post in Berlin, convinced that such a pack of bloodthirsty idiots could only lead Germany to disaster.

Bohr's Institute took on a less frivolous air, as it became a way station for fleeing exiles. Bohr himself became increasingly preoccupied with the task of finding them jobs in safe countries. And then, in the Christmas season of 1938, nuclear fission was discovered at the Kaiser Wilhelm Institute in Berlin. Within a matter of weeks, the wizards of the microworld knew that their days of happy innocence were at an end. A nuclear bomb might well be possible, and Nazi Germany appeared to have a head start.

As if to underscore their apprehensions, a lid of military secrecy was clamped on work in a new wing at the Kaiser Wilhelm, and the Nazi state seized the uranium-rich tailings of Czechoslovakia's radium mines. Refugee physicists pushed Great Britain and then the United States into the race for the bomb. In a world gone mad, quantum fields would have to wait.

By the time the physicists were free to return to their first love, the torch had been passed to America. The quantum theory of electromagnetism, known as *quantum electrodynamics* or QED, was completed in 1947 by two New Yorkers still in their twenties, Richard Feynman and Julian Schwinger. Since this theory served as a model for all later theories, we will introduce it at this point.

QUANTIZING THE FIELD

Feynman would have been in his element in the halcyon days of Bohr's Institute. Fond of highjinks and high living, he was known to jolt his mind out of a rut by working at a back table in a night club, inspired rather than distracted

Richard Feynman.

by the glare of stage lights and the blare of the sound system. Schwinger, on the other hand, preferred Einstein-like solitude. Working independently, they completed their theories within weeks of one another. As with Schrödinger and Heisenberg, the theories looked on the surface so different that it took a considerable effort to prove they were in fact the same.

The essence of Feynman's style is simplicity, so it is his version of the theory that we present here. Figure 19-1 shows the quantum version of how the electromagnetic field transfers energy and momentum between two electrons. Newton's continuous force is replaced by a "package" transfer in the form of a photon. The force law is replaced by a formula that gives the *probability that any given amount of momentum and energy will be transferred.*

This figure, known as a *Feynman diagram,* is more than a way to visualize the process. It contains an exact recipe for calculating the probability. The calculations are tedious and will not be discussed in detail, but once the right diagrams have been drawn they are so automatic that they can be turned over to a computer.

QED finally settles the old question of the nature of electric charge. It is nothing but *the ability to emit and absorb photons.* When one particle can emit or absorb another, the particles are said to be *coupled.* A quantitative measure of this ability is the *coupling strength,* which for electromagnetism can be expressed as

$$\alpha = \frac{e^2}{\hbar c}$$

FIGURE 19-1.

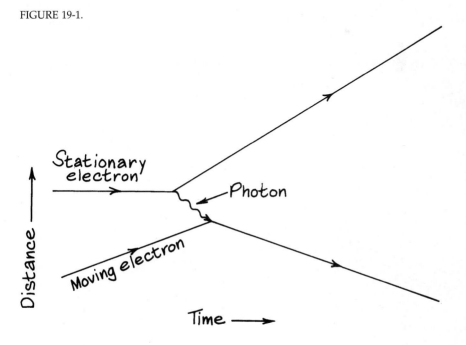

This is a pure number, independent of the choice of units, and its value (which is somewhat less that 1/100) determines the strength of the force. For each emission or absorption of a photon, the probability of the process is multiplied by this factor, so adding more photons to a diagram reduces the probability. For example, the diagram in Figure 19-2 shows an inelastic collision of two electrons, in which part of the collision energy is used to create a free photon. Because there is one more photon emission, this is roughly 100 times less probable than the elastic collision shown in Figure 19-1.

The exchange of a photon is of course a form of "quantum tunneling," because the electron has no internal source of energy from which it can create a photon. The energy and momentum must be "borrowed" for the short time that the photon is in flight. The uncertainty relations allow one to calculate the probability that this will happen. For the one-photon diagram in Figure 19-1, it turns out to be $1/p^4$, where p is the momentum transferred by the photon. This result is not terribly important, but we mention it because when a light object collides with a heavy one, p is proportional to the *sine of half the angle* through which it is deflected. Does that sound familiar? Rutherford's alpha-particle scattering law, which he derived from Newtonian orbit theory and Born obtained from wave mechanics pops up again, in an entirely different sort of model, yet another example of the hazards of model building.

In a collision on the nuclear scale, this exchange of a single photon is the dominant process. But electromagnetic fields can extend to vast distances because a photon has no rest mass and can have as little energy as we please. Long-range forces are transmitted by a nearly continuous flow of photons of negligible energy. On the scale of electrons moving in atoms, this is very nearly the case, which is why Schrödinger's hydrogen calculation worked. Feynman and Schwinger made only small corrections to the hydrogen energy levels. But

FIGURE 19-2. Inelastic electron collison (photon emitted).

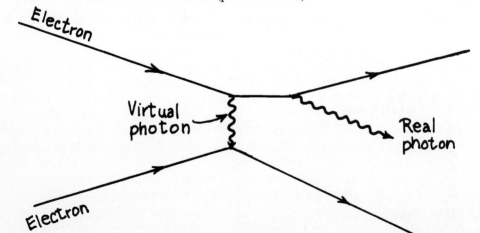

these were crucial in establishing the validity of their theory, for they had recently been measured.

Other fields are transmitted by quanta other than the photon. Quantum field theories rest on a completely reciprocal relationship between particles and fields:

> The nature of a field is completely determined by the properties of the particle that transmits it, while the nature of a particle depends solely on the ways in which it couples to fields.

After we have examined the full list of known fields, we will return to this rule to see how it operates in practice. Just as in general relativity, the field concept that began so modestly as a substitute for action at a distance materializes as matter itself!

Thus the task of subatomic physics is clear: find all the fundamental particles and their fields. But unfortunately, this turned out to be far more complicated than anyone in those bright heroic days of the 1920s or even the 1940s realized. The "zoo" of fundamental particles is much larger, and is filled with far more exotic specimens, than anyone had reason to suspect. And other fields are considerably more complicated than electromagnetism. Before we can go on, we must pause to deal with a few messy facts.

MATTER AND ANTIMATTER, FERMIONS AND BOSONS

One of the more interesting features of quantum field theory, discovered by the British theorist Paul Dirac in the 1920s, is that for each type of particle there must exist an antiparticle, opposite in electrical charge but equal in mass. Within a few years of the prediction, it was confirmed by the discovery of the *positron*, identical to the electron in all respects save that it carries a positive electric charge. It has since been confirmed for many other types of particle.

The matter-antimatter distinction is the basis for another division of particles into two classes, *fermions* and *bosons*. The electron is a fermion, while the photon is a boson. Particles with spin that is zero or an integer multiple of $\hbar$ are bosons, while half-integer spins belong to fermions.

A fermion can be created only if at the same time its own antiparticle is created. Similarly, it can be destroyed only if it encounters its own antiparticle, although it can be transformed into another kind of fermion. Bosons, however, can be freely created or destroyed as long as enough energy is available. Thus, fermions have a kind of permanence, and they serve as the building blocks of matter. Bosons are more ephemeral, and serve as the field quanta, the "glue" that holds matter together.

One of the great puzzles of cosmology is how the universe came to contain matter without an equal quantity of antimatter. We know that our corner of creation, meaning our galaxy and its near neighbors, consists entirely of matter, except for an occasional antiparticle produced in a collision. Nowhere in the universe do we find the kind of titanic explosions that happen when large hunks of matter and antimatter meet. In 1967 Andrei Sakharov suggested that certain irreversible processes that could have happened during the first instants of creation gave rise to a slight excess—less than a part in a billion—of matter over antimatter. Soon afterward, the rest of the matter and antimatter annihilated, and all the matter in the universe today comes from that tiny excess.

THE ATOM SMASHERS

It sometimes strikes even the physicists who work with them as ironic that some of the largest machines ever built are used to look at the smallest things we know about. The current generation of *particle accelerators* consists of machines whose dimensions are measured in miles, with price tags measured in billions of dollars. In effect, these machines are gigantic "microscopes."

In an ordinary microscope we see things by means of light, photons with energies of a few electronvolts. To see much smaller details, we need probes with much shorter wavelengths, and thus with far greater momentum and energy. The study of subatomic particles involves looking for details that are far smaller than the size of a nucleus, which is measured in *femtometers*, 10^{-15} meter. Today's most powerful accelerators can show details ten thousand times smaller than this, using protons or electrons with billions or trillions of electronvolts of energy.

The rest energy (henceforth we shall simply use the word mass, for we are now in the realm where energy units are used for mass) of a proton is 938 million electronvolts, while that of an electron is about a half million. Thus these particles must be pushed to speeds within an eyelash of that of light, making them hundreds or thousands of times heavier than at rest. When a particle is that heavy, only another particle as massive as itself is a suitable target, for a heavy object can transfer only a small portion of its momentum to a lighter one—if you punch a balloon, the momentum of your fist doesn't change very much. Thus the largest accelerators produce two beams of particles that collide head-on. There are only a handful of these machines in the entire world.

When particles collide, much of their energy is converted into rest energy of new and unstable particles. From a single collision of two particles, dozens of new ones emerge. They are carefully tracked by tens of thousands of electronic "eyes" wired to computers that reconstruct what happened in the collision. These arrangements are so complicated that teams of hundreds of scientists are required to build and operate them. A picture emerges slowly, from the study of billions of individual collisions. Much of the effort is concerned with simply identifying new and unstable particles that flash briefly into existence.

QUARKS AND LEPTONS

By the 1960s, the accelerators had revealed that the two particles that make up atomic nuclei, *protons* and *neutrons*, simply did not qualify as fundamental. They were fairly large, and seemed to have a fairly complex inner structure, like little "atoms within the atom." Murray Gell-Mann and George Zweig of Caltech proposed in 1963 that they were combinations of smaller fermions that, unlike any particle previously known, carried fractions of the fundamental unit of electric charge.

To emphasize the uniqueness of these particles, Gell-Mann chose the fanciful name *quarks*. This peculiar choice of terminology proved to be fateful. Following Gell-Mann's lead, particle theorists have tended to play a game of "one-upmanship" with terminology, resulting in a lexicon of "cute" names that tend to baffle (and sometimes outrage) nonspecialists. This practice has often proved a barrier to wider understanding, however much fun it may provide for insiders. Thus one of the more confusing and annoying tasks you will face in the rest of this chapter will be to wade through a lot of verbiage that is far from self-explanatory.

It takes two kinds of quarks to make protons and neutrons: *u*-quarks, which carry $+\frac{2}{3}$ units of electric charge, and *d*-quarks with charge $-\frac{1}{3}$. A proton consists of two *u*'s and a *d*, while a neutron is two *d*'s and one *u*.

The electron remains a fundamental particle, a member of a family called *leptons*. It is closely related to an electrically neutral object called an *electron neutrino*, represented by the symbol ν_e. These neutrinos play an important role in radioactivity, and in the nuclear fusion reactions that power the stars.

These two leptons and two quarks are pretty much all we need to make ordinary matter as we know it. Unfortunately, however, this basic pattern of four is repeated twice over. Each of the three charged fermions has two heavier, unstable versions. Each spontaneously breaks up into combinations of particles from the generation below it. Table 19-1 lists the complete roster of fundamental fermions, with their rest masses in millions of electronvolts (MeV). Each repeat of the fundamental quartet is called a *generation*. So far, none of the neutrinos has been shown to have a measurable mass, though it is unlikely that their masses are exactly zero.

TABLE 19.1 Fundamental Fermions

Leptons		*Quarks*	
-1	0	$-\frac{1}{3}$	$+\frac{2}{3}$
e	ν_e	*d*	*u*
0.51	< 0.00002	~ 7	~ 4
μ	ν_μ	*s*	*c*
106	< 0.25	~ 200	~ 1500
τ	ν_τ	*b*	*t*
1784	< 70	~ 4700	~ 176,000

The second and third generations have a fleeting existence in ordinary matter as "virtual" particles, and a few are produced when high energy radiation from space reaches our atmosphere, but otherwise they have little role in our universe today. They were, nonetheless, crucially important in the early moments of the big bang, when all fermions were equally abundant. Sakharov's process for circumventing the balance of matter and antimatter requires the participation of particles from all three generations. Without them, we could not be here today!

The existence of these six quarks and six leptons is something that no theory pretends to explain. For now, they must simply be taken as given. The letters d, u, s, c, b, t stand for *down, up, strange, charmed, bottom, and top.** The heavier leptons are the *muon* and *tau*, and their neutrinos the *mu-neutrino* and *tau-neutrino*.

The first of the unstable fermions, the muon, was discovered in 1938. Within ten years, it had become clear that it was exactly like an electron in all measurable properties, save that it was 200 times heavier. This provoked Isidore I. Rabi, a leader of the scientific community in the United States, to exclaim, "Who ordered that?" We now know that whoever it was ordered up a whole menu of such objects! The last one, the t-quark, was discovered in 1994.

It is important to note that each generation contains a neutrino, and neutrinos have so far proved to have little or no rest mass. In the early 1990s, an extended search for additional varieties of neutrino was conducted. It would have detected any neutrino with a mass of less than 45,000 MeV, but found only the three familiar kinds. If there is no fourth lightweight neutrino, it seems highly unlikely that there is a fourth generation. Thus at this level, the search for the building blocks of matter seems to be have come to an end. The knowledge of quarks was hard-won because, unlike any particles studied before, quarks are *never found alone*. Instead, they appear only in combinations, of which three kinds are allowed: *three quarks, three antiquarks,* or *one quark and one antiquark.* The quark model was developed by studying these composite particles and imagining how to build them up from a few smaller constituents. Because quarks have never been observed apart for other quarks and the fields that bind them together, the estimates of their masses given in Table 19-1 are only approximate.

The confusion created by the inseparability of quarks led even Gell-Mann to harbor doubts as to whether they really existed. Two decades of heroic efforts to blast loose free quarks in violent collisions, or to find them in trace amounts in ordinary matter, met only with failure. Their fractional charges raised a few eyebrows, because the combination rules all too neatly ensured that the particles built up out of quarks would always have whole-number charges, concealing the fractional nature of the quark charges. Ever since the debacle with the aether, whenever nature seems to be conspiring to hide something from experimental view scientists tend to "smell a rat" and begin to wonder whether it really exists. Nonetheless, a variety of reactions studied since the 1970s reveals the presence of these fractional charges.

---

*For some more whimsical theorists, b and t stand for *beauty* and *truth!*

And both the peculiar combination rules and the refusal of quarks to remain single turned out in the end to have a perfectly natural explanation. By the late 1970s, these clues led to an understanding of the strongest and most complicated field of all. This theory is called *QCD*, which stands for *Quantum Color Dynamics*.

THE SORCEROR'S MAGIC BROOM

The quanta of the QCD field are remarkably similar to photons. They are electrically neutral objects with zero rest mass, and are called *gluons*. In a moment we will see how appropriate the name is, for gluons "stick" to quarks quite tenaciously. The theory was formulated in 1976, and the existence of gluons was confirmed in 1979.

The QCD coupling, the property that is analogous to *charge* in electrodynamics, is called *color*. It does not, of course, have anything to do with color in the ordinary sense of the word. This term gained acceptance because of a neat analogy between the quark combination rules and the mixing of primary colors to obtain white.

Electric charge can be positive or negative. Color is similar, but has *three kinds each* of positive and negative, designated R, G, B, for red, green, and blue. The distinction is qualitative rather than quantitative; the force on a quark is the same whatever color it carries. The coupling strength of color, designated by the symbol α_s, is about $1/10$, so the gluon force is much stronger than electromagnetism, which is governed by the photon coupling strength α, more than ten times smaller.

Quarks carry positive colors, while antiquarks have negative colors ("anticolors"), designated by $\bar{R}, \bar{G}, \bar{B}$. Leptons carry no color at all, so it is the color coupling that makes quarks different from leptons.

Since opposite charges attract, electric forces always act in such a way as to keep a system neutral. The same is true for QCD, but because of the additional dimension of color there are *two* ways to achieve neutrality (keeping the color analogy, the term equivalent to "neutral" is "white"):

1. A color paired with its anticolor $(R\bar{R}, G\bar{G}, B\bar{B})$
2. An equal mix of three colors or anticolors $(RGB$ or $\overline{RGB})$

This is the origin of the two kinds of allowed quark combinations. The first combination is called a *meson*, the second a *baryon* or *antibaryon*. The proton and neutron are the lightest and most stable baryons—no other baryon lives more than a nanosecond. All mesons are unstable, since the pairing of matter with antimatter must eventually lead to annihilation.

There is a crucial distinction between photons and gluons, and it makes the QCD field very different in its action from electromagnetism. Photons carry no electric charge, but gluons do carry color. Each gluon, in fact, carries one color and one anticolor. It is this feature that is responsible for the inseparability of quarks.

Remember that electric charge is the ability to emit and absorb photons. *Color* is then *the ability to emit and absorb gluons.* Photons have a very passive role—one charged particle emits a photon, and another later absorbs it. But gluons have a more active role—one gluon can emit another gluon, because its color gives it the power to do so. Since photons have no electric charge, one photon cannot emit another.

Thus gluons proliferate in flight, adding strength to the force they transmit. This more than compensates for the natural tendency of fields to diminish with distance. If two quarks try to move apart, the force binding them actually gets *stronger,* and the coupling strength becomes larger.

A gluon can also turn into a quark-antiquark pair, since it already carries the requisite color and anticolor. In an undisturbed particle it has to "borrow" the energy to do this, and the virtual pair quickly reverts to gluons. But if energy is available from some external source such as the kinetic energy of a collision, the new quark and antiquark may become real, and a new particle is created.

Taken together, these effects explain why it has proved impossible to separate a single, isolated quark from its partners. An attempt to do that is illustrated in Figure 19-3. Protons are bombarded by high-energy electrons from a

FIGURE 19-3. An electron collides with a quark inside a photon.

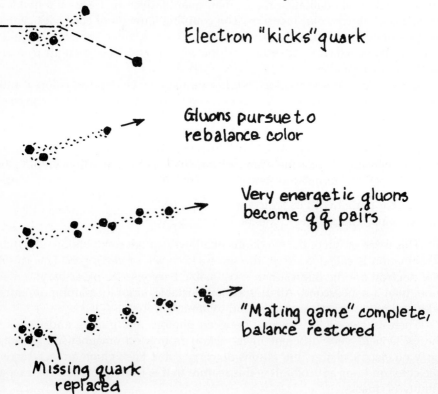

Electron "kicks" quark

Gluons pursue to rebalance color

Very energetic gluons become $q\bar{q}$ pairs

"Mating game" complete, balance restored

Missing quark replaced

particle accelerator. A quark in one proton is given a strong push by a close encounter with an electron. It starts to fly free, but cannot escape the pursuing gluons, moving at the speed of light. They proliferate in flight, deriving their energy from the fleeing quark.

Some of this energy goes into making quark-antiquark pairs, and the quark "mating game" begins. After a very short time, the struck quark is no longer "free," but has acquired an antiquark partner, to form a swift-moving meson. It is escorted by more mesons and possibly some baryon-antibaryon pairs. A quark has dropped into the proton, to fill the empty slot. All of these states are "white," so color balance is restored.

What emerges from the collision is a "jet" of particles, all moving in nearly the same direction. By summing their momenta and energies, experimenters can reconstruct the motion of the struck quark and the forces that acted on it.

Thus quark combinations are "indestructible" in a very funny sense. They are like the "magic broom" in the legend of *The Sorcerer's Apprentice*, which was portrayed in Walt Disney's celebrated animated feature film *Fantasia*. With his master away, an apprentice sorceror tries to save himself some labor by charming a broom to come to life and carry water. But he doesn't know the spells required to make it stop, and when he tries to do so by chopping it to bits with an ax, each fragment regenerates its missing parts and he winds up with a veritable army of water-toting brooms.

And so it is with quark combinations. Any attempt to break them up leads to the creation of *new partners*, assuring that no quark will long remain single.

A COSMIC RECYCLER

From the earliest beginnings of quantum field theory, it was known that there must be another field operating on the nuclear scale, in addition to the one that holds nuclei together. In terms of the modern quark picture, we know that this field transforms one kind of fermion into another.

For example a d-quark may be transformed into a u-quark by emitting an electron and an electron antineutrino. If this happens to one of the d-quarks inside a neutron, the particle is transformed into a proton, a process known as *nuclear beta decay*. The Feynman diagram for this reaction is depicted in Figure 19-4. The backward arrow on one of the lines is a convention marking it as an antiparticle, in this case an *anti-electron-neutrino*. This antiparticle must be emitted to maintain the matter/antimatter balance. A u may also transform into a d, changing a proton into a neutron. Because d-quarks are heavier than u's, energy must be supplied to drive this reaction. All the heavy charged fermions are disposed of by reactions of this sort, transformed into their first-generation counterparts.

The quanta that mediate these reactions come in electrically charged forms, the $W^{\pm}$, and a neutral form, the Z^0. Their masses are given in Table 19-2, which lists all the known and presumed fundamental bosons. The weak coupling treats all fermions equally.

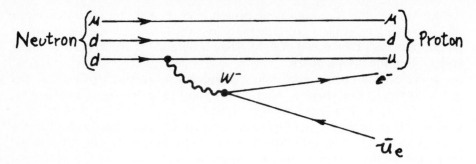

FIGURE 19-4. Neutron beta decay.

For historical reasons, this field is called the *weak interaction*. Today we know that it has the same coupling strength as electromagnetism. The reason it appears weaker is because the quanta that transmit it have very large rest masses. Thus it takes a considerable "minimum investment" to create one of these particles, which makes the tunneling probability very small. The uncertainty relation only allows them to exist for a very short time, so that any force they produce can only extend to a short range.

The role of the weak interaction in our universe is terribly important. First of all, it is the reason why nearly all the matter in our universe is made of first-generation fermions, because it is only through the weak interaction that one kind of fermion can be transformed into another. Second, it is essential to the reactions that power stars such as our Sun. These stars derive most of their energy from fusing four hydrogen nuclei into a helium nucleus. But most hydrogen nuclei consist of a lone, isolated proton, while a helium nucleus consists of two neutrons and two protons. Along the way, two protons must be transformed into neutrons, releasing two positrons and two electron neutrinos. Without the weak interaction, the Sun and other stars could not shine.

Neutrinos are not coupled to either photons or gluons: the weak interaction is their main link to the world. The tunneling probability for weak inter-

TABLE 19-2. Fundamental Bosons

Field	Particle	Mass	Spin	Couples to:
Electromagnetic	Photon	0	1	Charge
Strong nuclear	Gluon	0	1	Color
Weak nuclear	$W^\pm$	80,130	1	All
	Z^0	91,190	1	Fermions
Gravity	Graviton?	0	2	Energy
Higgs?	Higgs?	?	0	Rest mass

actions is so small that a neutrino can pass right through a nucleus with only a small chance of interacting. This makes neutrinos the most penetrating form of radiation known. Most of the neutrinos produced in the core of our Sun pass right through its outer layers and, when they reach the Earth, pass right through it, too. To monitor this process, a detector the size of a railroad car has been operating since 1970 deep in the Homestake Mine in South Dakota. Though more than a million billion neutrinos pass through the detector each second, only once every two or three days does one interact with a nucleus in the detector, leaving a trace of its existence.

One of the greatest puzzles in astrophysics is that far fewer neutrinos are detected than would be expected from the rate at which the Sun radiates energy. New and more sensitive detectors are being constructed in the hope of finding out why.

Two more bosons are required to round out the Standard Model. One of these is the *graviton*. Gravity seems to us like a strong force, because we feel the cumulative effects of the attraction of the huge mass of the Earth. But in atomic terms, gravity is terribly feeble. The gravitational attraction of an electron to a nucleus is 10^{41} times weaker than the electrical attraction. For this reason, no one has yet succeeded in detecting an individual graviton. But since the force of gravity is well understood on the macroscopic level, we know exactly what a graviton must be like. Reconciling this picture of gravity with Einstein's geometric theory has proved a mathematically difficult task, but some progress has been made.

Finally there is an empty slot waiting for a particle called the *Higgs boson*, after theorist Peter Higgs, who first suggested its existence. Its role is to create rest mass, for those particles that have it. Unfortunately, the theory of this field is so incomplete that we know little about its detailed properties, and no particle that might be a Higgs boson has yet been found.

PARTICLES AND FIELDS

We are now equipped to examine in detail the mutual interdependence of particles and fields. Starting with the fermions, we look back to Table 19-1, and consider the ν_e, the lightest and simplest. It couples only to the weak bosons and, like everything else, to the graviton. This gives it its elusive character. But add *electric charge −1*, and it would become an *e*, a stable constituent of an ordinary atom. Give it instead *color*, and you have a *u*-quark. Each of these particles is differentiated from its second- and third-generation counterparts by a different coupling to the Higgs field.

Bosons carry integer units of spin, indicated in Table 19-2, and the value of the spin plays a major role in determining the character of the force it transmits. If the spin is an even number, as is the case with the graviton, the force will always be attractive. When it is odd, as with the photon or gluon, the force can either attract or repel.

Thus electricity is an inverse-square force because the photon has no rest mass. It can either attract or repel because the photon has one unit of spin. All the wondrous phenomena of electromagnetism expressed in Maxwell's theory derive from the starkly simple properties of the photon.

Still, it must be stressed that just like the geometric picture of fields in general relativity, quantum field theory remains an incomplete scheme. We know next to nothing about the Higgs field, so until the right sort of boson is actually discovered it remains a conjecture. Speculations about a field responsible for generating spin are even more fanciful. A theory called *superstrings*, dating from the early 1980s, purports to unify quantum fields with Einstein's geometric scheme and eliminate *all* the loose ends in both theories, so its promulgators have dared to call it the *Theory of Everything*. But formidable mathematical difficulties prevent it from coming up with testable predictions, so for the time being it remains a theory of *nothing*.

THE QUANTUM FOAM

One of the obvious questions, in the wake of the triumph of the Standard Model, is how big is a point? Do quarks and leptons have smaller parts, on a scale smaller than the limit of detail revealed by our present accelerators, which is 10^{-19} meter? Will there ever be an end to this game?

In the early days of the quantum theory, Max Planck reflected on what it could mean if his constant h joined the speed of light c and Newton's gravitational constant G as one of the fundamental constants of nature. He noted that there was one way to combine these three constants to obtain a length,

$$r_p = \sqrt{\frac{G\hbar}{c^3}} = 1.6 \times 10^{-35} \text{ meter}$$

now called the *Planck length*. We have taken the modern liberty of replacing h with $\hbar$. This length was so small that he speculated it might be a fundamental minimum "unit" of length. In the 1960s, John Wheeler resurrected this speculation and interpreted it in the light of general relativity and the uncertainty relations.

The uncertainty relations say that we cannot be sure how much energy (and therefore mass) is contained within a small region of space-time. The smaller we make the region, the larger the uncertainty. In a sphere whose radius is the Planck length, it is about 20 micrograms, which is just enough to make the sphere a black hole. The energy-time uncertainty relation allows it to exist for no more than 10^{-43} second, the time equivalent of the Planck length. Wheeler pointed out that on this scale space can no longer be regarded as smooth and continuous, for it is filled with these tiny virtual black holes rapidly popping in and out of existence. It becomes what he calls a *quantum foam*. Thus Planck's speculation was on the right track: though the Planck length is not exactly a "unit" of length, it probably represents the smallest meaningful size for anything.

The smallness of the Planck length must give us pause. There is plenty of room for more Chinese boxes between the 10^{-19} scale we can now study and this ultimate limit. But at least we can be reassured that the game must eventually come to an end.

One trend that may prove significant is that as particles move closer together, the differences between fields tend to diminish. Move close enough to an electron, and α gets larger, while close to a quark α_s gets smaller. Eventually, somewhere approaching the Planck length, the two become equal. Since it is only their couplings that distinguish one particle or field from another, on the smallest scale all particles and fields become indistinguishable. This holds out hope that they are really different guises for one fundamental kind of field, the kind of unified field theory that Einstein sought in vain for the last twenty-five years of his life.

AN ACCIDENTAL UNIVERSE?

To close this book, let us keep a promise made at the end of Chapter 12. What can we learn by putting the Standard Model into our picture of the big bang?

The first answer is something about the fields themselves. In the theory as it stands, the rest masses of particles must be inserted as experimental numbers that cannot be explained. The dream of a field that generates its own matter remains unrealized. One school of thought holds that this can be achieved only through the unified field that must have dominated the first instants of the big bang, when particles were so crowded together that gravity was as strong as the gluon field.

As the universe expanded, the individual fields "condensed out." Some particles that were originally massless acquired mass, and each field thus acquired its individual character. The subtle interplay of fields in this process left its mark in the pattern of masses, which can be read as a "recording" of the early moments of the big bang.

A question comes immediately to mind: Did the pattern of masses arise out of logical necessity, or was it simply random?

There are hints in both directions; on the one hand, our present incomplete field theory does not leave the choice of masses entirely free. In the weak interaction the masses of the $W^{\pm}$ and Z^0 are closely related. If a more complete theory finds enough such connections and extends them to quarks, perhaps God really did have no choice after all.

On the other hand John Wheeler insists that during the early moments of the big bang, the universe was so small that quantum fluctuations must have played a major role. He would expect particle masses to be random, but also to have just those lucky values that allowed *us* to evolve. This is called the *anthropic principle*. Since we know that we are here, but otherwise know little about why the universe is the way it is, let us examine how things had to be in order to allow us to eventually come upon the scene. For example, it can be shown that if the fundamental unit of charge is changed by even a fraction of 1 percent, a stable star like our Sun becomes impossible.

This viewpoint takes on a fascinating dimension if we imagine that our universe was born as a fluctuation, a mere fleck in the cosmic foam of some larger entity that we might call a *superuniverse*. Most such fluctuations would last only the characteristic Planck time. But an occasional terribly rare one might deviate from the norm by just enough to cross a threshold that enables it to tap some source of energy and grow rapidly, a process called *inflation*. Such might be the origin of our universe.

What can happen once can happen again and again, so this conjecture suggests that ours is not the only universe. Furthermore, we can imagine that each universe comes with its own toss of the cosmic dice. The particle masses and with them the structure of matter go up for grabs. The process repeats in endless variety, each universe as unique as a snowflake. In the unending span of space and time, every imaginable universe will get its turn!

Herein lie the seeds of a possible synthesis of the rival visions of Bohr and Einstein. God does have choices—an infinity of them—and is obliged to try them all. There may be reason enough here to satisfy Einstein's yearnings, yet disorder enough to delight Bohr's sense of mystery.

How and when (if ever) such outrageous ideas will touch base with reality is anyone's guess. Each generation of scientists finally comes to the shores of its continent of solid fact. For the time being the ocean beyond can be crossed only in the imagination. This has always been the driving wheel of scientific creativity. The thrill of holding such visions in one's mind is one of the sweetest rewards of the calling of scientist.

Such heady wine calls for a note of caution, and so it is appropriate to close this work with the sage words of a celebrated American philosopher and psychologist, William James:

> I am convinced that the desire to formulate truths is a virulent disease. It has contracted an alliance lately in me with a feverish personal ambition, which I never had before, and which I recognize as unholy in such a connexion. I actually dread to die until I have settled the Universe's hash in one more book! . . . Childish idiot—as if formulas about the Universe could ruffle its majesty, and as if the common sense world and its duties were not eternally the really real!

Summary

The last step in the development of quantum physics was to quantize fields. Newton's continuous momentum transfer is replaced by a "package" of energy and momentum via the tunneling process. In the case of the electromagnetic field, the package is a virtual photon. The world of quantum fields is one in which matter consists of particles that are point-like field sources held together by the transfer of such virtual particles. The role of a particular particle in this scheme depends on whether it is a fermion or a boson. Fermions are the building blocks, while bosons transmit the forces. Subatomic physics is complicated by the existence of several kinds of each. There are two twelve-member families of fundamental fermions, leptons (the electron and its relatives) and quarks (subunits of the particles found in the nucleus). Most of these are unstable and play no major role in ordinary matter. Several types of boson are required to account for all the known forces. Gravity has not yet been fully integrated into this scheme.

Afterword

To be human is to wonder. Children wonder for a while, before we teach them to be smug about the obvious and to stop asking silly questions. It is easier to pay someone to retain a little of the child and do our wondering for us. We then take comfort in the assumption that anyone devoted to such esoteric pursuits must be insensitive, perhaps even inhuman. With our artists, we perform the equal disservice of regarding them as *too* sensitive.

Occasionally we are given a glimpse of the finished product. The baby is displayed behind glass, well-scrubbed, and one need not know about the delivery room (it is soundproofed). Thus we are spared the agony of wonder, which is not unlike love and makes as little (or as much) sense as love. But wonder is just too human to fully repress, and it does turn up elsewhere. Some of us turn to fads for the occult, which interpreted by our twentieth-century minds becomes a cartoon science. More often, we find ourselves left with nothing to wonder about (or to love) but what remains of ourselves after the loss of yet another portion of our humanity.

I, for one, refuse to believe that nothing can be done about this empty place, or about the more general disease of which it is but a minor symptom. But as long as we are sundered so, let me remain one of the children and wonder.

R. H. M.

Bibliography

CLASSICAL PHYSICS

Dijksterhuis, E. J.: *The Mechanization of the World Picture* (Oxford University Press, London, 1961). A comprehensive history of mechanics up to Newton's time, including ancient and medieval thought.

Galilei, Galileo: *Two New Sciences* (Macmillan, New York, 1914, available in Dover paperback). Galileo's dialogues, elegantly translated by Henry Crew and Alfonso De Salvio.

Koestler, Arthur: *The Watershed* (Doubleday, Garden City, NY, 1960). A heroic biography of Kepler by an author with a somewhat mystical theory of scientific creativity.

Mach, Ernst: *The Science of Mechanics* (Open Court, La Salle, IL, 1960). A historical and philosophical analysis by the leading champion of positivism in physics.

Newton, Isaac: *Principia* (University of California Press, Berkeley, 1962, two volumes). Difficult at points because of the archaic style, but the ideas are presented clearly.

RELATIVITY

Gardner, Martin: *Relativity for the Millions* (Cardinal, New York, 1965). Possibly the best popular exposition of relativity.

Pais, Abraham: *Subtle Is the Lord* (Oxford University Press, Oxford and New York, 1982). The definitive biography of Einstein, written by a physicist, with an emphasis on his scientific achievements.

Sciama, Dennis: *The Physical Foundations of General Relativity* (Doubleday, Garden City, NY, 1969). A short, readable introduction to the general theory with a stress on its relation to cosmology.

Taylor, E. F., and J. A. Wheeler: *Spacetime Physics* (Freeman, San Francisco, 1963). The text on relativity for physics undergraduates. Though some familiarity with the calculus is required, it is beautifully written and treats a large number of examples.

QUANTUM THEORY

Blaedel, Niels (translated by Geoffrey French): *Harmony and Unity* (Science Tech Publishers, Madison WI, and Springer-Verlag, Berlin, 1988). A Danish biography of Niels Bohr that places his science and his world view in their cultural context.

Bohr, Niels: *Atomic Physics and Human Knowledge* (Vintage, New York, 1966). Bohr's views on the broader implications of his physical thinking.

Cline, Barbara: *Men Who Made a New Physics* (Signet, New York, 1966). Biographies of all the leading figures in the development of the quantum theory, with some exposition of their ideas.

Forman, Paul: *Weimar Culture, Causality, and Quantum Theory 1918–1927: Adaptations by German Physicists to a Hostile Intellectual Environment, in Historical Studies in the Physical Sciences,* vol. III (University of Pennsylvania Press, Philadelphia, 1971). The title is self-explanatory.

Heisenberg, Werner (translated by Arnold Pomerans): *Physics and Beyond* (Harper & Row, New York, 1971). Autobiographical and philosophical essays by a leading figure of the quantum revolution.

Moore, Walter: Schrödinger: *Life and Thought* (Cambridge University Press, Cambridge, England, 1989). A scientific biography that stresses the cultural influences on and philosophical views of its protagonist.

Toulmin, Stephen, and June Goodfield: *The Architecture of Matter* (Harper, New York, 1962). An excellent history of atomism and other ideas related to the structure of matter, from ancient to modern times.

OF GENERAL INTEREST

Harrison, Edward: *Masks of the Universe* (Macmillan, New York, 1985). A history of cosmological ideas from antiquity to the present, in the context of cultural anthropology, with an exposition of the underlying science.

Holton, Gerald: *Thematic Origins of Scientific Thought: Kepler to Einstein* (Harvard University Press, Cambridge, MA, 1973). Thoughtful, scholarly, but readable essays on Kepler, Bohr, and Einstein, with an emphasis on the cultural and psychological roots of their ideas.

Kuhn, Thomas S.: *The Structure of Scientific Revolutions,* 2d ed. (University of Chicago Press, Chicago, 1970). A modern treatise on a philosophy of science that emphasizes historical factors.

Zukav, Gary: *The Dancing Wu Li Masters* (Morrow, New York, 1979). An exposition of twentieth century physics and its compatability with oriental philosophy, by a writer affiliated with the "physics and consciousness" school.

Questions and Exercises

It is customary, in conventional physics courses, to equate understanding with the ability to calculate. While this book is predicated on the assumption that these are independent achievements, working a few numerical exercises can help clarify the meaning of the principles involved.

The exercises in this section were devised for use in an organized course, with an instructor to show you examples and otherwise assist you in learning how to work them. If you are using this book for self-study, you may find it difficult to work many of the exercises. Those marked with an asterisk (*) require the use of formulas given in footnotes in the text, and tend to be somewhat more difficult than the others. Those marked with a dagger (†) require considerable computational or analytic skill, and are intended to challenge the mathematically adept.

The questions are designed to stimulate your thinking about nonquantitative aspects of the concepts introduced in the text.

CHAPTER ONE

Questions

A. Give an example of two objects in which the lighter one falls faster than the heavier one.
B. Give an argument against Galileo's claim that a body rolling down an inclined plane is not different in kind from one falling freely.

Hints

- For motion at constant speed, or in calculating average speed, use the formula $x = vt$, where x, v, t, are distance, speed, and time.
- In all exercises involving acceleration, assume the acceleration is uniform.
- The simplest exercise on accelerated motion involve comparing velocity v_1 at the start of the acceleration to v_2 at the end of a time t. The relevant formula is then $at = v_2 - v_1$.

- To simplify arithmetic, in all problems that involve the acceleration due to gravity, use the value $g = 10$ m/s^2.
- The formula $x = \frac{1}{2}at^2$ applies *only* to accelerated motion starting from rest.

Exercises

1. An airliner flies 2040 miles in 4 hours. What is its average speed, in miles per hour?
2. A sprinter does the 100-meter dash in 10.0 seconds.
   a What is his average speed?
   b If he could maintain this pace for a mile (1610 meters), what would his time for the mile be?
3. A sports car can accelerate at 5 m/s^2. How long will it take to reach a speed of 30 m/s (about 68 mph) from a standing start?
4. A car speeds up from 22 m/s to 28 m/s in 3 seconds. What is its acceleration, in m/s^2?
5. A car moving at 48 m/s brakes to a stop in 6 seconds. What is its acceleration?
*6. How far does a freely falling body drop in 3 seconds?
*7. How far can the sports car in Exercise 3 travel in 4 seconds, starting from rest?
*8. How long does it take an object dropped from atop a 320-meter-high tower to reach the ground?
9. A jet plane starts from rest on the runway and accelerates uniformly. It takes off 25 seconds later, having rolled 1000 meters. Find:
   a Its average speed
   b Its speed at takeoff
   c Its acceleration
10. A car brakes uniformly to a halt in 5 seconds, traveling 100 meters from the moment the brakes were applied.
   a What was its average speed while braking?
   *b What was its acceleration?
   c What speed was it going at the moment the brakes were applied?

CHAPTER TWO

Questions

A. It has been observed that the more languages a person can speak, the easier it is to learn another language. Does this argue for or against the existence of a "Principle of Superposition" for this kind of learning?
B. Propose a principle analogous to a conservation law for a field outside the natural sciences (it need not be *true*).
C. Explain how Galileo's analysis of projectile motion supports his forbidden argument that the Earth can be moving without our sensing it.
D. Cite at least three everyday examples that illustrate momentum conservation.

Hints

- In exercises on momentum conservation, find the *sum* of the momenta of the two objects, $m_1v_1 + m_2v_2$. In most cases only one is moving. The sum of momenta must be the same after the collision.

- If you are told the objects *stick together*, treat them as *one object* with a mass equal to the *sum* of their masses.
- If you are given the velocity of one of the objects *after* the collision, calculate its momentum and subtract it from the momentum before the collision to find the momentum of the other object.
- Don't forget—motions in opposite directions have *opposite signs* of momentum!
- An elastic collision is one in which the *difference* in velocity of the two objects changes sign but has the same magnitude after the collision as before.

Exercises

1. A ball is dropped from a moving train, and is observed by one observer on the train and another on the ground.
   a   Sketch the path of the ball, as seen by each observer.
   b   Repeat these sketches, assuming the train was *speeding up* when the ball was dropped.
2. A car drives off a sheer vertical cliff moving horizontally at 30 m/s. It strikes the ground 2 seconds later.
   a   How far from the base of the cliff does the car land?
   b   How high is the cliff?
3. A baseball is thrown with a horizontal velocity component of 30 m/s and a vertical component of 20 m/s.
   a   How long is it in flight?
   b   How far does it travel?
4. A football is punted with a "hang time" (time in flight) of 6 seconds, and is caught 54 meters from where it was kicked.
   a   What was its horizontal velocity component?
   b   What was its vertical component?
   *c   How high was it at the top of its trajectory?
*5. A baseball leaves the bat with a velocity of 30 m/s, moving upward at 45 degrees to the horizontal. How far does it travel?
†*6. Using the trigonometric relationship shown in the footnote on the range of a projectile, prove that:
   a   The maximum range for a projectile with a given velocity comes when it is projected at a 45-degree angle.
   b   For any range shorter than this maximum range, there are *two* values of θ that will give that range.
7. A car is on a ferryboat that is stationary on the water. The car weighs 2 tons, while the ferryboat weighs 10 tons. The car is driven at a speed of 12 mph with respect to the deck for a distance of 60 feet, and then stops.
   a   While the car is in motion, how fast is the ferryboat moving?
   b   When the car stops, does the ferryboat also stop?
   c   After the car stops, how far has the ferryboat moved?
8. A clay ball of mass 1 kg and velocity of 6 m/s strikes a stationary ball of mass 2 kg. After the collision the balls are stuck together and moving at 2 m/s. Show that momentum is conserved in the collision.
9. A clay ball of mass 3 kg and velocity of 16 m/s strikes a stationary ball of mass 5 kg and sticks to it.
   a   What was the momentum of the 3-kg ball *before* the collision?
   b   What was the momentum of the 5-kg ball *before* the collision?

   c What was the combined momentum of the balls *after* the collision?

   d What was the *velocity* of the balls after the collision?

   e What is the momentum of the 3-kg ball *after* the collision?

   f How much momentum was *transferred* between the balls during the collision?

10. A ball of mass 2 kg and velocity 10 m/s strikes a stationary ball of mass 1 kg. After the collision the 2 kg ball is moving at 4 m/s in the same direction it was going before the collision.

   a What was the momentum of the 2-kg ball *before* the collision?

   b How much momentum was *transferred* between the balls during the collision?

   c What is the velocity of the 1-kg ball after the collision?

   d Show that this was *not* a perfectly elastic collision.

11. A 50-kg figure skater moving 10 m/s meets her 70-kg partner moving 5 m/s in the opposite direction and they hang on to one another. What is their speed after they meet?

†12. A 1-kg object moving at 9 m/s collides *elastically* with a stationary 2-kg object. What are the velocities of the two objects after the collision? (*Hint*: You will need to set up two equations to solve for these two unknowns.)

CHAPTER THREE

Questions

A. The normal method of comparing masses is to weigh them on a balance. What assumptions must be made in order for this to be a valid procedure?

B. Explain why it is possible for a roller coaster to go upside down in a vertical loop without falling off the track.

C. Explain the operation of a rocket in terms of Newton's laws.

Hints

- Force is *defined* as momentum transfer divided by time. In most cases, this is equal to the mass times the acceleration.
- For circular motion in a circle of radius r at speed v, the acceleration is $a = v^2/r$.

Exercises

1. In a collision, two objects are in contact for $\frac{1}{10}$ second and 50 kg-m/s of momentum is transferred. What is the average *force* acting in the collision?

2. A force of 10 N acts on an object of mass 5 kg. What is its acceleration?

3. What is the force on an object of mass 3 kg accelerating at 7 m/s²?

4. a Convert your own mass to kilograms (1 kg = 2.2 lb).

   b Calculate the *force of gravity* on your body, in newtons.

5. A car that can normally accelerate at 4 m/s² is towing an identical car. What is the fastest it can accelerate?

6. A car is moving at 15 m/s around a curve of radius 75 meters. What is its acceleration?

7. A tire will skid if subjected to a sideways acceleration of more than 4 m/s². What is the fastest speed at which a car equipped with such tires can round an unbanked curve of radius 100 meters without skidding?

8. What force must be exerted to keep a 6-kg mass moving in a circle of radius 10 meters at a speed of 20 m/s?

†9. A roller coaster executes a complete vertical loop of radius 10 meters. What is the minimum speed it must have at the top of the loop in order to not fall off the track?

†10. A rocket exhaust spews out 100 kg of mass moving at 200 m/s each second. What is the force exerted on the rocket?

11. A car of mass 1000 kg brakes uniformly to a stop from 20 m/s in 4 s. What is the force exerted on the car?

12. The velocity due to the Earth's rotation of an object on the equator is about 400 m/s and the radius of the Earth is 6.4×10^6 m. Calculate the acceleration of this object due to the Earth's rotation.

CHAPTER FOUR

Questions

A. The four satellites of Jupiter discovered by Galileo obey Kepler's laws. Going down Table 4-1, indicate which aspects of Newton's Law of Gravity are supported by this observation.

B. Although the gravitational constant was unknown in Newton's time, it was still possible to compare the mass of the Earth to that of the Sun. Explain.

C. One of Galileo's arguments in favor of the Copernican theory was that Venus had phases like the Moon. Explain how this is was relevant. Could Tycho's scheme also explain this observation?

Hints

• If the long axis a of a planetary orbit is measured in units of the a of the Earth's orbit, and the period is measured in Earth years, Kepler's third law becomes an equality: $T^2 = a^3$.

• The constant G in Newton's Law of Gravity is 6.67×10^{-11} for distances measured in meters, masses in kilograms, and force in newtons.

Exercises

Planet	Major Axis	Orbital Period
Mercury	0.39	0.24
Venus	0.72	0.61
Earth	1.00	1.00
Mars	1.52	1.90
Jupiter	5.20	12.0

1. The table above gives the length of the major axis, in units of the Earth's, and the orbital period in Earth years, for the five innermost planets. Verify that these satisfy Kepler's third law.

2. The orbital velocity of the Earth is about 30,000 m/s and the distance from the Sun is about 1.5×10^{11} meters.

    a Assuming the orbit is a circle, calculate the Earth's orbital acceleration.

    †b The acceleration due to gravity is 9.9 m/s² at the Earth's surface, and the Earth's radius is 6,400,000 meters. Use these values and the result of (a) to estimate the mass of the Sun, in units of the Earth's mass.

  3. Calculate the force between two objects of mass 10 and 100 kg, separated by 0.2 meter.

  †4. Show that for *circular* orbits, Kepler's third law implies that the *accelerations* of the planets are proportional to the inverse-square of their distances from the sun.

  †5. An Earth satellite in low orbit ($r = 6500$ km) goes around the Earth in 80 minutes. What is the orbital radius for a *geosynchronous* satellite that goes around once in 24 hours, thus appearing to remain stationary as seen from Earth?

  6. Show that the acceleration due to gravity at the surface of the Earth ($r = 6.4 \times 10^6$ m and $M = 6.0 \times 10^{24}$ kg) has the expected value (around 9.9 m/s²).

CHAPTER FIVE

Questions

A. Cite examples of "holistic" and "reductionist" thinking in a field outside the natural sciences.

B. A baseball is hit a long distance, slowed by air resistance, and then caught. Detail the energy conversions that take place from the moment it leaves the bat.

C. Explain why more work must be done to accelerate a car from 60 mph to 70 mph than to accelerate it from 0 mph to 10 mph.

Hints

- The *work* done by a force is always equal to the force multiplied by the motion in the direction of the force, $Fx \cos \theta$. If the force is unopposed, all this work goes into increasing the *kinetic energy* of the object.
- If a force is opposed by *friction*, the work done against friction becomes heat. This will not be all the energy—that is only true if the friction is equal to the propulsive force. Friction is usually *opposite* the direction of motion ($\cos \theta = -1$).
- Work done in *raising* an object of mass *m* to a height *h* goes into gravitational potential energy, *mgh*. For simplicity use $g = 10$ m/s² for the acceleration due to gravity.
- For motion at constant speed, power is force times velocity.

Exercises

1. How much work is done by a force of 10 N on an object that moves 100 meters:
  a In the same direction as the force
  b Perpendicular to the force

2. How much work is done in lifting a 5-kg mass a height of 30 meters?

3. A force of 5 N pushes on an initially stationary object of mass 4 kg for a distance of 10 meters in the direction of the force.
  a How much work is done?
  b What is the final speed of the object?

4. A force of 50 N acts on an object, opposed by a frictional force of 10 N, over a distance of 100 meters.

   a How much work is done by the 50-N force?
   b How much of this work goes into increased kinetic energy?
5. A cyclist climbs a hill 10 meters high on a road 100 meters long, at a constant speed of 5 m/s. The combined mass of bike and rider is 70 kg, and the force exerted to propel the bike is 100 N.
   a How much *work* is done by the force?
   b What is the increase in *gravitational potential energy*?
   c What happens to the rest of the energy?
   d How much *power* (in *watts*) must the cyclist generate? Convert this to *horsepower*.
6. A car moving at steady 20 m/s is propelled by a force of 150 N. What is the required power, in watts and in horsepower?
7. A ball dropped on a hard floor bounces back to $\frac{4}{9}$ of the height it was dropped from. Calculate:
   a The fraction of the ball's energy that is lost
   b The ratio between the ball's speed just after it leaves the floor to its speed just before hitting the floor
8. A cyclist approaches a hill moving at 10 m/s. If she coasts without pedaling, what is the maximum height she can reach?
9. Refer back to the example of an elastic collision in Chapter 2 and verify that the *kinetic energy* is unchanged in the collision.
†10. Show that in an elastic collision of a moving object with one of equal mass at rest, the moving object stops and the other continues at the original speed of the moving object.
11. In a 30-day month, how many *kilowatt-hours* of electrical energy must be used to keep a 50-watt light bulb burning continuously?
†*12. **a** Using the footnote formula for a spherical gravitational field, find the potential energy of a 1-kg mass at the Earth's surface ($r = 6.4 \times 10^6$ m and $M = 6.0 \times 10^{24}$ kg).
   b Find the *velocity* at which the kinetic energy of this mass would be equal in magnitude to your answer to (a) (this is the so-called escape velocity).

CHAPTER SIX

Questions

   A. Explain how the concept of field makes potential energy "less mysterious."
   B. A charged object moving parallel to a magnetic line of force will experience no force at all. Explain why.
   C. A steady magnetic field does no work on a charge. Explain why.
   D. Though he worked only on the electromagnetic field, Maxwell in effect proved that *any* field whose action is not instantaneous, but which propagates at a finite speed, must be capable of producing some form of radiation. Explain why.

Exercise

1. Gravity can also be regarded as a field. In this case, the field strength is the *force per unit mass*.
   a From Newton's Law of Gravity, write a formula for gravitational field strength.
   b To what other physical quantity is gravitational field strength identical?

CHAPTER SEVEN

Questions

A. Give three examples of wave phenomena not mentioned in the text.

B. An *octave* in music is a doubling of the frequency. Which harmonic of a music string is one octave above the fundamental?

C. Explain how "fingering" a stringed instrument changes the pitch of the note sounded.

D. Radio stations are frequently required to avoid broadcasting in certain directions in order not to interfere with faraway stations. One way to do this is to broadcast from two towers exactly one-half wavelength apart. Explain why this works, and make a sketch indicating the directions in which the signal from such a station would be strong or weak.

E. An audiophile mistakenly hooks up a stereo so that the leads to one speaker are reversed from the way they should be. How should that affect the sound of the system?

Hints

- The speed of light is 300,000,000 (3×10^8) m/s. The speed of sound in air varies a bit with temperature, but is in the vicinity of 340 m/s.
- Wave interference is determined by the *difference* between the distances traveled by the waves, measured in wavelengths: $(x_1 - x_2)/\lambda$. If this quantity is an *integer* the waves reinforce perfectly, while if it is a *half integer* they cancel.

Exercises

1. An FM station broadcasts at 100 MHz. What is the length of the waves it transmits?

2. Humans can typically hear sounds of frequencies from 20 Hz to 17,000 Hz. What are the corresponding wavelengths?

3. What are the wavelengths of the first three harmonics on a music string that is 0.3 meter long?

4. A radio station broadcasts from two antennas spaced one-half wavelength apart. On a sketch, show the location of the towers and draw lines heading out from them along which the signals *cancel*, and another line along which they *reinforce* perfectly.

5. Two speakers are placed 3 meters apart and are sounding a note of wavelength 2 meters. A listener walks along a line 4 meters in front of the speakers.
   a Show that the waves from the two speakers will cancel at the points along the line that are directly in front of the speakers.
   †b In all, at how many points along the line will the waves reinforce?

6. Lightning strikes a point 3 km from you.
   a How long does it take the light to reach you? Is that a noticeable delay?
   b How long does it take the sound of the lightning (the thunderclap) to reach you? Is that a noticeable delay?

7. A piano string of length 1 m sounds a fundamental frequency of 440 Hz (A above middle C).
   a What is the wavelength of these waves on the string?
   b Find the speed of waves on the string.

†8. Waves travel around a hoop of radius R. Derive a formula that gives the allowed wavelengths of standing waves on the hoop.

CHAPTER EIGHT

Questions

A. Explain why it is necessary to rotate the Michelson-Morley apparatus.
B. Describe what Michelson *expected* to see as he looked through the eyepiece with the apparatus in rotation.
C. Suppose a Michelson experiment were performed with sound rather than light, with a microphone connected to headphones in place of the eyepiece. The experiment is performed outdoors on a day when a steady wind is blowing. Would there be an effect? What would the observer hear as the apparatus rotated?
D. One early critic of the Michelson-Morley experiment pointed out that they should have repeated the experiment several times at different seasons of the year. Can you guess why?

Hints

* The exact formula for the Lorentz factor γ should be used whenever the speed is a appreciable fraction (greater that $\frac{1}{10}$) of the speed of light. When it is much less than the speed of light, the approximate formula is more appropriate.
* In most exercises on relativity you will be given the velocity as a fraction of the speed of light, e.g., $v = 0.5c$. In these cases you already have v/c to use in the formula for γ—you need not divide by c.

Exercises

1. Calculate γ for $v = 0.8c$.
2. Calculate γ for $v = 1.5 \times 10^8$ m/s.
3. Calculate γ for $v = 0.002c$.
4. A ferryboat that moves at 5 mph crosses a river that is 2 miles wide and moves at 3 mph, reaching a point on the other side directly opposite its starting point. How far does the ferryboat travel *in the water*?
†5. In a Michelson interferometer with two paths of equal length L, the expected difference in path lengths due to the Earth's motion was $L(\gamma^2 - \gamma)$. In the Cleveland instrument, $L = 10$ meters and the wavelength of light employed was around 0.5 μm (5×10^{-7} meter). Show that from the Earth's orbital motion, 30,000 m/s, Michelson-expected the interference bands to shift by $\frac{1}{5}$ of their separation.

CHAPTER NINE

Questions

A. Devise a procedure for synchronizing clocks at two ends of a train by means of light signals, and show that an observer who regards the train as moving will *not* consider them synchronized.

B. Does the postulate of relativity require *gravity*, as well as light, to have a finite velocity of propagation?

C. In the train example, which observers believe that the distance between the telegraph poles is equal to the length of the train? In what reference frame do they believe this to be true?

D. In the example of the spaceship with mirrors, state the observable fact on which both observers agree.

E. Show that the assumption that there are two different velocities both of which satisfy the postulate of relativity leads to a logical contradiction.

CHAPTER TEN

Questions

A. Explain one sense in which the Lorentz contraction is "real," and another sense in which it is "not real."

B. Criticize the following statement: "In relativity, there is no one reality, only the mutually irreconcilable realities of different observers."

C. Critique the following statement: "Time is now regarded as simply a fourth dimension of space, in no way different from the other three."

Hints

- To convert the length of an object in its rest frame to its length in a frame in which it is moving, *divide* by γ.
- To convert time elapsed on a clock to time elapsed in a frame in which the clock is moving, *multiply* by γ.
- To simplify your arithmetic in Exercise 3, use $c = 300$ m/μs (meters per microsecond).

Exercises

1. If an observer moving at $0.6c$ with respect to you reports that 10 minutes elapsed on her clock, how much time would you estimate elapsed on your clock?
2. What is the length of an object moving at $0.6c$, if its rest length if 10 meters?
3. Assume the train in Chapter 9 has rest length 1000 meters and is moving at $0.6c$. According to observers on the ground, the conductor at the rear of the train makes his observation of the telegraph pole 2.5 μs (0.0000025 second) before the conductor at the front observes his.
   a How far does the train move between the two observations?
   b Use the Lorentz factor to obtain the length of the train while moving.
   c Add your answers to (a) and (b) to get the distance between the poles, according to the observers on the ground.
   d Multiply the time interval between observations by c.
   e Subtract the square of your answer to (d) from the square of your answer to (c) and take the square root. This is a four-dimensional invariant and should equal the rest length of the train.
†4. By calculating the difference in the times it took light to reach the two ends of the train in Exercise 3, verify that the time interval was in fact 2.5 μs.

5. An artificial Earth satellite orbiting at a speed of 7500 m/s carries a precision clock.
   a Calculate the difference between γ and 1.0 for this speed.
   b How much time will this clock lose in one day (86,400 seconds) due to the moving-clock effect? (We will see in Chapter 12 that there is another relativistic phenomenon that also affects this clock.)

†6. If the astronaut in the twin paradox example tries to follow events at home by listening to radio broadcasts from Earth, the transmission that reaches him just as he arrives at the star will have been broadcast a bit less than 2 days after his departure.
   a Show that this is true in the *Earth* reference frame.
   b Show that this is true in the *astronaut's* reference frame.

7. In the spaceship example, if Joe replies immediately to Sue's first message, calculate what her clock reads when she receives it.

8. Repeat the calculation in the spaceship example for $v = 0.8c$, showing that both observers agree that Joe's clock reads 30 min when he receives Sue's communication.

9. Two spaceships are heading toward one another at 0.8c. Find their relative velocity in the rest frame of one of the ships.

10. Show that in the frame in which both Joe's and Sue's spaceships are traveling with equal and opposite velocities, the velocity is $\frac{1}{3}c$.

CHAPTER ELEVEN

Questions

A. Discuss how relativity has modified the laws of energy conservation and momentum conservation.

B. For each of the processes listed below, indicate whether the mass *increases*, *decreases*, or *remains the same*.
   (1) An automobile battery is charged.
   (2) A hot steel bar is allowed to cool down.
   (3) A rubber band is stretched.
   (4) Hydrogen and oxygen burn to form water in a sealed, insulated container.
   (5) Repeat (4) in a container that allows heat to escape.
   (6) Two atoms bound together in a molecule are separated.

C. State whether two observers in relative motion in spaceships will agree or disagree about each of the following quantities:
   (1) Their relative speed
   (2) The mass of either ship
   (3) The rate at which a clock on the other ship runs
   (4) The length of the other ship
   (5) The width of the other ship
   (6) The velocity of light
   (7) The velocity of an object moving inside one of the ships

Hints

• The energy equivalent of a kilogram of mass is 9×10^{16} joules. One kilowatt-hour (a few cents' worth) of electricity is 3.6 million joules.

Exercises

1. What is the mass of an object with 3 kg rest mass if it is moving at $v = 0.8c$?
2. How fast must an object move in order to have a mass double that when it is standing still?
†3. Using the approximate formula for γ, show that the total energy $\gamma m_0 c^2$ of a moving object is simply the sum of its rest energy and the prerelativistic form of kinetic energy.
4. The fastest interplanetary rockets have attained speeds of 60 km/s. If such a rocket has a rest mass of 1000 kg, how much greater is its mass when moving at full speed?
5. A large power plant produces 15 million kilowatt-hours of electricity per day, enough to power a medium-sized city.
   a Convert this amount of energy to joules.
   b What is the *mass* of this much energy?
6. The Tevatron is a particle accelerator that produces a beam of protons moving at $v = 0.99999946c$. How many times more massive are these protons than they would be at rest?
7. It takes 2.25 million joules of energy to boil 1 kg of water. How much more mass does the steam have than the water it was boiled from?
8. The power of nuclear weapons is measured in *kilotons* of TNT equivalent. One kiloton is 4×10^{12} joules. How much rest mass is converted to energy in a 900-kiloton bomb?

CHAPTER TWELVE

Questions

A. Explain the relationship between Galileo's falling body law and the Principle of Equivalence.
B. Explain why Newton's instantaneous action at a distance is not admissible in a relativistic theory.
C. Describe in words why one twin in the twin paradox ends up younger than his brother by describing how time passes on Earth and on the ship during the voyage, from the point of view of each of the twins.
D. Describe the *space-time track* of an orbiting Earth satellite.
E. The expanding universe should look the same no matter where you are in it. Explain why this is so.
F. Indicate what ultimate end *you* would prefer for the universe, and why you feel that way.

Hints

• Exercises 3 and 4 use the formula for the gravitational time effect in a constant field, ah/c^2.

Exercises

1. With a world globe and a string, verify that the shortest path from Chicago to Rome does not follow the east-west line on which they both lie.

2. How long does it take light to travel 30 km? How far would a falling body drop in this time?

3. A precision clock is placed on top of a mountain 4500 m high. How much would this clock gain in 100,000 seconds (a bit more than a day), compared to a clock at sea level?

4. The clock in Exercise 5 of Chapter 10 is orbiting at an altitude of 300 km. How much time does it gain in 100,000 s from the gravitational time effect?

*†5. Compare the rate of a clock on the surface of the Sun ($r = 700,000$ km, $m = 2 \times 10^{30}$ kg) to one at the surface of the Earth ($r = 6400$ km, $m = 6 \times 10^{24}$ kg, orbit radius 150,000,000 km, orbital velocity 30 m/s), taking into account the Earth's motion and the gravity of both the Earth and the Sun. Newton's constant $G = 6.67 \times 10^{-11}$ for mass in kilograms and distances in meters.

*†6. A neutron star has a mass of 3×10^{30} kg and a radius of 10 km. Calculate the gravitation time shift at its surface.

CHAPTER THIRTEEN

Questions

A. Describe the following phenomena in terms of the atomic theory:
   (1) Boiling of a liquid
   (2) Surface tension in a liquid
   (3) The fact that when a gas is compressed by a piston, it gets hotter

B. Give arguments supporting Thompson's conclusion that what he had discovered was a very light object carrying a single unit of charge, rather than a heavy one carrying many units.

C. Cite what you consider the three most convincing arguments for the atomic nature of matter.

Exercise

1. Look up a table of atomic weights, and give the recipes by weight for forming the following compounds:
   a   Hydrogen chloride (HCl)
   b   Carbon dioxide (CO_2)
   c   Sulfuric acid (H_2SO_4)

2. The boiling point of hydrogen is $-253°$ Celsius. How many degrees is that above absolute zero?

3. Since gas temperature is simple a measure of the average kinetic energy of molecules, what is the ratio between the *speeds* of hydrogen and oxygen molecules (H_2 and O_2) at the same temperature?

CHAPTER FOURTEEN

Questions

A. The behavior of atoms in solids and liquids suggests that they act pretty much like hard spheres. Explain how this is possible using the plum-pudding model and the planetary model.

B. Explain why Rutherford considered it *especially* significant that at least a few of the alphas Marsden observed actually came off backward.
C. In a Wilson cloud chamber, alpha particles leave a trail of water droplets and in most cases this trail is straight, though a few tracks exhibit a "kink" at which the direction changes suddenly. Use this as evidence to support Rutherford's nuclear hypothesis.

Exercises

1. The diameter of a gold atom is about 0.25 nanometer (a nanometer is 10^{-9} meter). Marsden used gold foil about 1 μm (10^{-6} m) thick. How many gold atoms did each alpha particle pass through?
†2. In order to be deflected backward, an alpha particle had to pass within 2×10^{-14} meter of the nucleus. Imagine each nucleus to be the center of a "target" this size and estimate the probability that the alpha would be deflected backward in passing through the gold foil of Exercise 1.
3. If in a Geiger-Marsden experiment 1200 alphas per hour are observed at 30 degrees from the beam, how many will be seen at 90 degrees? At 150 degrees?
4. The constant v_0 in the Balmer formula is 3.3×10^{15} Hz. Find the frequency of he spectrum line corresponding to $n = 1, m = 2$.

CHAPTER FIFTEEN

Questions

A. The particle features are more evident when using x-rays rather than visible light. Explain why.
B. Which two Bohr orbits are farthest apart in energy?
C. List all the physical principles used in Bohr's model of hydrogen, and indicate which of these were:
(1) Carried over from Newtonian physics
(2) Taken from the earlier quantum theory of Planck and Einstein
(3) New ideas original to Bohr

Hints

- For some of these exercises, you will need the relation between frequency and wavelength, $\lambda v = c$.
- Planck's constant $h = 4.14 \times 10^{-15}$ eV-s.

Exercises

1. What is the energy, in eV, of one photon of light of frequency 10^{15} Hz?
2. The human eye can see light between wavelengths of 0.4 and 0.7 μm. What is the range of *photon energies* the eye can see?
3. Calculate the wavelength of a gamma-ray photon with an energy of 1 billion eV.
4. What is the frequency of light given off when the electron in a hydrogen atom drops from:
a The $n = 2$ state to the ground state
b The $n = 3$ state to the $n = 2$ state

5. In a photoelectric experiment, the threshold frequency (the lowest light frequency that can eject electrons) is 1.5×10^{15} Hz.
   a How much energy (in eV) must be provided to liberate an electron?
   b If light of frequency 3.5×10^{15} Hz is used, what is the maximum energy of the emitted electrons?
6. How much energy (in eV) is required in a hydrogen atom in order to:
   a Lift an electron from the ground state to the $n = 2$ state
   b Completely remove an electron from a ground state atom
7. There is a subatomic particle called a *muon* that is identical to an electron in most respects but is 207 times heavier. If a muon replaced the electron in a hydrogen atom, what would be the ground state energy E_1?

CHAPTER SIXTEEN

Questions

A. As you move from low Bohr orbits to high ones, do the wavelengths get longer or shorter?
B. If a photon and electron have the same kinetic energy, which has the shorter wavelength?
C. List the arbitrary or paradoxical aspects of Bohr's model that are eliminated in the Schrödinger picture. Do any remain?
D. Does wave mechanics shed light on the significance of Planck's constant? Explain.

Hints

- Planck's constant in conventional units is $h = 6.6 \times 10^{-34}$ joule-second.
- At velocities below 0.1c, ignore relativistic effects and use the formulas $p = mv$ and $K = \frac{1}{2}mv^2$ for momentum and kinetic energy.

Exercises

1. Show that the de Broglie formula gives the right wavelength for a photon, given that the energy carried by light is its momentum multiplied by c.
†2. Show algebraically that an electron always has a wavelength that is shorter than that of a photon of equal energy (kinetic energy).
3. The electrons in Davisson's experiments typically had velocities of around 10^7 m/s. Given that the electron mass is 9×10^{-31} kg, calculate the wavelengths of these electrons. (The answer is comparable to the spacing of atoms in crystals.)
4. Rutherford's alpha particles had mass 6.6×10^{-27} kg and velocity $v = 0.03c$. Calculate their wavelengths. (The answer is comparable to the sizes of nuclei.)

CHAPTER SEVENTEEN

Questions

A. If Planck's constant were smaller, would that make uncertainty more or less significant?

B. Critique the following statement: "The uncertainty relations show that there is always a limit to how accurately one can measure something."

C. Explain why the heavier an object is, the less important is uncertainty in limiting predictions of its future.

D. Some of the most recently discovered subatomic particles show a large spread of values when repeated measurements are made of their masses. How can this be interpreted in the light of the uncertainty relations?

Hints

- For problems in which *energy* is involved, use Planck's constant in eV-s. Where momentum is given, use the value for conventional units.

Exercises

1. An excited state of a nucleus has a lifetime of $\delta t = 10^{-18}$ second. What is its uncertainty in energy, in eV?

2. Use the formula $E = h\nu$ to convert the energy-time uncertainty relation to one relating uncertainties in *frequency* and time. Can you see any intuitive significance in the result?

*3. How far in the future can predictions be made at the scale of a small molecule (mass = 10^{-25} kg, size = 10^{-9} meter)?

CHAPTER EIGHTEEN

Questions

A. Which interpretation of the "Schrödinger's Cat" parable do you consider best (you can make up your own, if you wish)? Explain why you favor this one.

B. Do you feel the debate over the interpretation of the quantum theory is a worthwhile activity? Explain why or why not.

CHAPTER NINETEEN

Questions

A. Explain why it is regarded as desirable that the ultimate constituents of matter have no size at all.

B. If no simpler lever of reality underlies that of the Standard Model, would you still consider reductionism a valid approach to science? Defend your point of view.

C. What field coupling is shared by all fermions? By all particles whether bosons or fermions?

D. List all the fields to which each of the following fermions is coupled: mu-neutrino, electron, *t*-quark.

Exercises

1. What is the electric charge of each of the quark combinations listed below, and what class of particle (i.e., meson, baryon, antibaryon) is each?

    a uss
    b $u\bar{b}$
    c $\overline{uuu}$

2. The Λ^0 is the baryon uds, and may be thought of as a neutron with one d-quark replaced by an s. The neutron mass is 939 MeV. Using the masses in Table 19-1, estimate the mass of the Λ^0. (The actual mass, 1116 MeV, is somewhat lower than this estimate because higher-mass quarks tend to bind more tightly.)

3. The μ^- lepton is unstable. It is transformed into a ν_μ by the weak interaction $\mu^- \rightarrow \nu_\mu + e^- + \bar{\nu}_e$.
    a What virtual boson mediates this reaction?
    b Draw the Feynman diagram.

4. How many different kinds of *meson* quark combination are there?

†5. How many *baryon* combinations are there? (*Hint*: Consider separately baryons with three identical quarks, two identical, and all quarks different.)

Answers to Odd-Numbered Exercises

CHAPTER ONE

1. 510 mph
3. 6 seconds
5. 8 m/s^2
7. 40 meters
9. **(a)** 40 m/s   **(b)** 80 m/s   **(c)** 3.2 m/s^2

CHAPTER TWO

3. **(a)** 4 s   **(b)** 120 m
5. 90 meters
7. **(a)** 2 mph   **(b)** yes   **(c)** 10 ft
9. **(a)** 48 kg-m/s   **(b)** 0 kg-m/s   **(c)** 48 kg-m/s
   (d) 6 m/s   **(e)** 18 kg-m/s   **(f)** 30 kg-m/s
11. 1.25 m/s

CHAPTER THREE

1. 500 newtons
3. 21 newtons
5. 2 m/s^2
7. 20 m/s
9. 10 m/s
11. 5000 newtons

CHAPTER FOUR

3. 1.67×10^{-6} newton
5. 45,000 km

CHAPTER FIVE

1. (a) 1000 joules   **(b)** 0 joules
3. (a) 50 joules   **(b)** 5 m/s
5. (a) 10,000 joules   **(b)** 7000 joules
   (c) frictional heat   **(d)** 500 watts, or $\frac{2}{3}$ horsepower
7. (a) $\frac{5}{9}$ **(b)** $\frac{2}{3}$
11. 36 kWh

CHAPTER SIX

1. (a) $G\dfrac{M}{r^2}$ **(b)** acceleration due to gravity

CHAPTER SEVEN

1. 3 meters
3. 0.6 meter, 0.3 meter, 0.2 meter
5. (b) 3
7. (a) 2 m   **(b)** 880 m/s

CHAPTER EIGHT

1. 5/3 or 1.667
3. 1.000002

CHAPTER TEN

1. 12.5 minutes
3. (a) 450 meters   **(b)** 800 meters   **(c)** 1250 meters
   (d) 750 meters   **(e)** $1250^2 - 750^2 = 1000^2$
5. (a) 3.125×10^{-10} **(b)** 27 μs (2.7×10^{-5} second)
7. 40 minutes
9. $0.976c$

CHAPTER ELEVEN

1. 5 kg
5. (a) 5.4×10^{13} joules   **(b)** 0.6 gram (0.0006 kg)
7. 2.5×10^{-11} kg,

CHAPTER TWELVE

3. 50 ns (5×10^{-8} second)
5. Sun's gravity − 2.1×10^{-6}, Earth faster
Earth's gravity − 7×10^{-10}, Earth slower
Earth's motion − 5×10^{-9}, Earth slower

CHAPTER THIRTEEN

1. **(a)** 1:35   **(b)** 12:32   **(c)** 2:32:64
3. 4:1

CHAPTER FOURTEEN

1. 4000
3. 22/hr and 6/hr

CHAPTER FIFTEEN

1. 4.14 eV
3. 1.2×10^{-15} m
5. **(a)** 6.2 eV   **(b)** 8.3 eV
7. 2800 eV

CHAPTER SIXTEEN

3. 7.3×10^{-11} meter

CHAPTER SEVENTEEN

1. 4140 eV
3. 2 ns (2×10^{-9} s)

CHAPTER NINETEEN

1. **(a)** 0, baryon   **(b)** +1, meson   **(c)** −2, antibaryon
3. **(a)** W^-
5. 56 (6 with 3 identical, 30 with 2 identical, 20 all different)

Index